T0357879

THE
GENETIC
BOOK
OF THE
DEAD

RICHARD DAWKINS

THE GENETIC BOOK OF THE DEAD

A DARWINIAN REVERIE

ILLUSTRATED BY JANA LENZOVÁ

Yale UNIVERSITY PRESS

New Haven and London

First published in 2024 in the United States by Yale University Press
and in the United Kingdom by Head of Zeus Ltd,
part of Bloomsbury Publishing Plc.

Yale University Press books may be purchased in quantity for educational,
business, or promotional use. For information, please e-mail sales.press@yale.edu
(U.S. office) or sales@yaleup.co.uk (U.K. office).

9 7 5 3 2 4 6 8

A catalogue record for this book is available from the British Library.

Library of Congress Control Number: 2024930182
ISBN 978-0-300-27809-5 (hardcover)

Printed and bound in Germany by Mohn Media

FSC
www.fsc.org

MIX
Paper | Supporting
responsible forestry
FSC® C011124

Mike Cullen (1927–2001)
In grateful memory

You would have a problem with your research. You knew exactly where to go for help, and there he would be for you. I see the scene as yesterday. The wiry, boyish figure in the red sweater, slightly hunched like a spring wound up with intense intellectual energy, sometimes rocking back and forth with concentration. The deeply intelligent eyes, understanding what you meant even before the words came out. The back of the envelope to aid explanation, the occasionally sceptical, quizzical tilt of the eyebrows, under the untidy hair. Then he would have to rush off – he always rushed everywhere – and he would seize his biscuit tin by its wire handles, and disappear. But next morning the answer to your problem would arrive, in Mike's small, distinctive handwriting, two pages, often some algebra, diagrams, a key reference to the literature, perhaps an apt classical quotation or a verse of his own composition. Always encouragement.

We may know other scientists as intelligent as Mike Cullen – though not many. We may know other scientists who were as generous in support – though vanishingly few. But I declare, we have known nobody who had so much to give, combined with so much generosity in giving it.

From my eulogy at his memorial service
in Wadham College Chapel, Nov 2001

Contents

1

Reading the Animal

You are a book, an unfinished work of literature, an archive of descriptive history. Your body and your genome can be read as a comprehensive dossier on a succession of colourful worlds long vanished, worlds that surrounded your ancestors long gone: a genetic book of the dead. This truth applies to every animal, plant, fungus, bacterium, and archaean but, in order to avoid tiresome repetition, I shall sometimes treat all living creatures as honorary animals. In the same spirit, I treasure a remark by John Maynard Smith when we were together being shown around the Panama jungle by one of the Smithsonian scientists working there: 'What a pleasure to listen to a man who really loves his animals.' The 'animals' in question were palm trees.

From the animal's point of view, the genetic book of the dead can also be seen as a predictor of the future, following the reasonable assumption that the future will not be too different from the past. A third way to say it is that the animal, including its genome, embodies a *model* of past environments, a model that it uses to, in effect, predict the future and so succeed in the game of Darwinism, which is the game of survival and reproduction, or, more precisely, gene survival. The animal's genome makes a bet that the future will not be too different from the pasts that its ancestors successfully negotiated.

I said that an animal can be read as a book about past worlds, the worlds of its ancestors. Why didn't I use the present tense: read the animal as a description of the environment in which it itself lives? It can indeed be read in that way. But (with reservations to be discussed) every aspect of an animal's survival machinery was bequeathed via its genes by ancestral natural selection. So, when we read the animal, we are actually reading *past* environments. That is why my title includes 'the dead'. We are talking about reconstructing ancient worlds in which successive ancestors, now long dead, survived to pass on the genes that shape the way we modern animals are. At present it is a difficult undertaking, but a scientist of the future, presented with a hitherto unknown animal, will be able to read its body, and its genes, as a detailed description of the environments in which its ancestors lived.

I shall have frequent recourse to my imagined Scientist Of the Future, confronted with the body of a hitherto unknown animal and tasked with reading it. For brevity, since I'll need to mention her often, I shall use her initials, SOF. This distantly resonates with the Greek *sophos*, meaning 'wise' or 'clever', as in 'philosophy', 'sophisticated', etc. In order to avoid ungainly pronoun constructions, and as a courtesy, I arbitrarily assume SOF to be female. If I happened to be a female author, I'd reciprocate.

This genetic book of the dead, this 'readout' from the animal and its genes, this richly coded description of ancestral environments, must necessarily be a *palimpsest*. Ancient documents will be partially over-written by superimposed scripts laid down in later times. A palimpsest is defined by the *Oxford English Dictionary* as 'a manuscript in which later writing has been superimposed on earlier (effaced) writing'. A dear colleague, the late Bill Hamilton, had the engaging habit of writing postcards as palimpsests, using different-coloured inks to reduce confusion. His sister Dr Mary Bliss kindly lent me this example.

Besides his card being a nicely colourful palimpsest, it is fitting to use it because Professor Hamilton is widely regarded as the most distinguished Darwinian of his generation. Robert Trivers, mourning his death, said, 'He had the most subtle, multi-layered mind I have ever encountered. What he said often had double and even triple meanings so that, while the rest of us speak and think in single notes, he thought in chords.' Or should that be palimpsests? Anyway, I like to think he would have enjoyed the idea of evolutionary palimpsests. And, indeed, of the genetic book of the dead itself.

Both Bill's postcards and my evolution palimpsests depart from the strict dictionary definition: earlier writings are not irretrievably effaced. In the genetic book of the dead, they are partially overwritten, still there to be read, albeit we must peer 'through a glass darkly', or through a thicket of later writings. The environments described by the genetic book of the dead run the gamut from ancient Precambrian seas, via all intermediates through the mega-years to very recent. Presumably some kind of weighting balances modern scripts versus ancient ones. I don't think it follows a simple formula like the Koranic rule for handling internal contradictions – new always trumps old. I'll return to this in Chapter 3.

If you want to succeed in the world you have to predict, or behave as if predicting, what will happen next. All sensible prediction must be based on the past, and much sensible prediction is statistical rather than absolute. Sometimes the prediction is cognitive – 'I foresee that if I fall over that cliff (seize that snake by its rattling tail, eat those tempting *belladonna* berries), it is likely that I will suffer or die in consequence.' We humans are accustomed to predictions of that cognitive kind, but they are not the predictions I have in mind. I shall be more concerned with unconscious, statistical 'as-if' predictions of what might affect an animal's future chances of surviving and passing on copies of its genes.

This horned lizard of the Mojave, whose skin is tinted and patterned to resemble sand and small stones, embodies a prediction, by its genes, that it would find itself born (well, hatched) into a desert. Equivalently, a zoologist presented with the lizard could *read* its skin as a vivid *description* of the sand and stones of the desert environment in which its ancestors lived. And now here's my central message. Much more than skin deep, the whole body through and through, its very warp and woof, every organ, every cell and biochemical process, every smidgen of any animal, including its genome, can be read as describing ancestral worlds. In the lizard's case it will no doubt spin the same desert yarn as the skin. 'Desert' will be written into every reach of the animal, plus a whole lot more information about its ancestral past, information far exceeding what is available to present-day science.

The lizard burst out of the egg endowed with a genetic prediction that it would find itself in a sun-parched world of sand and pebbles. If it were to violate its genetic prediction, say by straying from the desert onto a golf green, a passing raptor would soon pick it off. Or if the world itself changed, such that its genetic predictions turned out to be wrong, it would also likely be doomed. All useful prediction relies on the future being approximately the same as the past, at least in a statistical sense. A world of continual mad caprice, an environmental bedlam that changed randomly and undependably, would render prediction impossible and put survival in jeopardy. Fortunately, the world is conservative, and genes can safely bet on any given place carrying on pretty much as before. On those occasions when it doesn't – say after a catastrophic flood or volcanic eruption or, as in the case of the dinosaurs' tragic end when an asteroid-strike ravaged the world – all predictions are wrong, all bets are off, and whole groups of animals go extinct. More usually, we aren't dealing with such major catastrophes: not huge swathes of the animal kingdom being wiped out at a stroke, but only those variant individuals whose predictions are slightly wrong, or slightly more wrong than those of competitors within their own species. That is natural selection.

The top scripts of the palimpsest are so recent that they are of a special kind, written during the animal's own lifetime. The genes' description of ancestral worlds is overlain by modifications and detailed refinements scripted since the animal was born – modifications written or rewritten by the animal's *learning* from experience; or by the remarkable memory of past diseases laid down by the immune system; or by physiological acclimatisation, to altitude, say; or even by simulations in imagination of possible future outcomes. These recent palimpsest scripts are not handed down by the genes (though the equipment needed to write them is), but they still amount to information from the past, called into service to predict the future. It's just that it's the very recent past, the past enclosed within the animal's own lifetime. Chapter 7 is about those parts of the palimpsest that were scribbled in since the animal was born.

There is also an even more recent sense in which an animal's brain sets up a dynamic model of the immediately fluctuating environment, predicting moment to moment changes in real time. Writing this on the Cornish coast, I take envious pleasure in the gulls as they surf the wind battering the cliffs of the Lizard peninsula. The wings, tail, and even head angle of each bird sensitively adjust themselves to the changing gusts and updraughts. Imagine that SOF, our zoologist of the future, implants radio-linked electrodes in a flying gull's brain. She could obtain a readout of the gull's muscle-adjustments, which would translate into a running commentary, in real time, on the whirling eddies of the wind: a predictive model in the brain that sensitively fine-tunes the bird's flight surfaces so as to carry it into the next split second.

I said that an animal is not only a description of the past, not just a prediction of the future, but also a *model*. What is a model? A contour map is a model of a country, a model from which you can reconstruct the landscape and navigate its byways. So too is a list of zeros and ones in a computer, being a digitised rendering of the map, perhaps including information tied to it: local population size, crops grown, dominant religions, and so on. As an engineer might understand the word, any two systems are 'models' of each other if their behaviour shares the same underlying mathematics. You can wire up an electronic model of a pendulum. The periodicity of both pendulum and electronic oscillator are governed by the same equation. It's just that the symbols in the equation don't stand for the same things. A mathematician could treat either of them, together with the relevant equation written on paper, as a 'model' of any of the others. Weather forecasters construct a dynamic computer model of the world's weather, continually updated by information from strategically placed thermometers, barometers, anemometers, and nowadays above all, satellites. The model is run on into the future to construct a forecast for any chosen region of the world.

Sense organs do not faithfully project a movie of the outer world into a little cinema in the brain. The brain constructs a virtual reality (VR) model of the real world outside, a model that is continuously updated via the sense organs. Just as weather forecasters run their computer model of the world's weather into the future, so every animal does the same thing from second to second with its own world model, in order to guide its next action. Each species sets up its own world model, which takes a form useful for the species' way of life, useful for making vital predictions of how to survive. The model must be very different from species to species. The model in the head of a swallow or a bat must approximate a three-dimensional, aerial world of fast-moving targets. It may not matter that the model is updated by nerve impulses from the eyes in the one case, from the ears in the other. Nerve impulses are nerve impulses are nerve impulses, whatever their origin. A squirrel's brain must run a VR model similar to that of a squirrel monkey. Both have to navigate a three-dimensional maze of tree trunks and branches. A cow's model is simpler and closer to two dimensions. A frog doesn't model a scene as we would understand the word. The frog's eye largely confines itself to reporting small moving objects to the brain. Such a report typically initiates a stereotyped sequence of events: turning towards the object, hopping to get nearer, and finally shooting the tongue towards the target. The eye's wiring-up embodies a prediction that, were the frog to shoot out its tongue in the indicated direction, it would be likely to hit food.

My Cornish grandfather was employed by the Marconi company in its pioneering days to teach the principles of radio to young engineers entering the company. Among his teaching aids was a clothesline that he waggled as a model of sound waves – or radio waves, for the same model applied to both, and that's the point. Any complicated pattern of waves – sound waves, radio waves, or even sea waves at a pinch – can be broken down into component sine waves – 'Fourier analysis', named after the French mathematician Joseph Fourier (1768–1830). These in turn can be summed again to reconstitute the original complex wave (Fourier synthesis). To demonstrate this, Grandfather attached his clothesline to rotating wheels. When only one wheel turned, the rope executed serpentine undulations approximating a sine wave. When a coupled wheel rotated at the same time, the rope's snaking waves became more complex. The sum of the sine waves was an elementary but vivid demonstration of the Fourier principle. Grandfather's snaking rope was a model of a radio wave travelling from transmitter to receiver. Or of a sound wave entering the ear: a compound wave upon which the brain presumably performs something equivalent to Fourier analysis when it unravels, for example, a pattern even as complex as whispered speech plus intrusive coughing against the background of an orchestral concert. Amazingly, the human ear, well, actually, the human brain, can pick out here an oboe, there a French horn, from the compound waveform of the whole orchestra.

Today's equivalent of my grandfather would use a computer screen instead of a clothesline, displaying first a simple sine wave, then another sine wave of different frequency, then adding the two together to generate a more complex wiggly line, and so on. The following is a picture of the sound waveform – high-frequency air pressure changes – when I uttered a single English word. If you knew how to analyse it, the numerical data embodied in (a much-expanded image of) the picture would yield a readout of what I said. In fact, it would require a great deal of mathematical wizardry and computer power for you to decipher it. But let the same wiggly line be the groove in which an old-fashioned gramophone needle sits. The resulting waves of changing air pressure would bombard your eardrums and be transduced to pulse patterns in nerve cells connected to your brain. Your brain would then without difficulty, in real time, perform the necessary mathematical wizardry to recognise the spoken word 'sisters'.

| S | I | S | T | ER | S |

Our sound-processing brain software effortlessly recognises the spoken word, but our sight-processing software has extreme difficulty deciphering it when confronted with a wavy line on paper, on a computer screen, or with the numbers that composed that wavy line. Nevertheless, all the information is contained in the numbers, no matter how they are represented. To decipher it, we'd need to do the mathematics explicitly with the aid of a high-speed computer, and it would be a difficult calculation. Yet our brains find it a doddle if presented with the same data in the form of sound waves. This is a parable to drive home the point – pivotal to my purpose, which is why I said it twice – that some parts of an animal are hugely harder to 'read' than others. The patterning on our Mojave lizard's back was easy: equivalent to *hearing* 'sisters'. Obviously, this animal's ancestors survived in a stony desert. But let us not shrink from the difficult readings – the cellular chemistry of the liver, say. That might be difficult in the same way as *seeing* the waveform of 'sisters' on an oscilloscope screen is difficult. But nothing negates the main point, which is that the information, however hard to decipher, is lurking within. The genetic book of the dead may turn out to be as inscrutable as Linear A or the Indus Valley script. But the information, I believe, is all there.

The pattern to the right is a QR code. It contains a concealed message that your human eye cannot read. But your smartphone can instantly decipher it and reveal a line from my favourite poet. The genetic book of the dead is a palimpsest of messages about ancestral worlds, concealed in an animal's body and genome. Like QR codes, they mostly cannot be read by the naked eye, but zoologists of the future, armed with advanced computers and other tools of their day, will read them.

To repeat the central point, when we examine an animal there are some cases – the Mojave horned lizard is one – where we can instantly read the embodied description of its ancestral environment, just as our auditory system can instantly decipher the spoken word 'sisters'. Chapter 2 examines animals who have their ancestral environments almost literally painted on their backs. But mostly we must resort to more indirect and difficult methods in order to extract our readout. Later chapters feel their way towards possible ways of doing this. But in most cases the techniques are not yet properly developed, especially those that involve reading genomes. Part of my purpose is to inspire mathematicians, computer scientists, molecular geneticists, and others better qualified than I am, to develop such methods.

At the outset I need to dispel five possible misunderstandings of the main title, *Genetic Book of the Dead*. First is the disappointing revelation that I am deferring the task of deciphering much of the book of the dead to the sciences of the future. Nothing much I can do about that. Second, there is little connection, other than a poetic resonance, with the Egyptian Books of the Dead. These were instruction manuals buried with the dead, to help them navigate their way to immortality. An animal's genome is an instruction manual telling the animal how to navigate through the world, in such a way as to pass the manual (not the body) on into the indefinite future, if not actual immortality.

Third, my title might be misunderstood to be about the fascinating subject of Ancient DNA. The DNA of the long dead – well, not *very* long, unfortunately – is in some cases available to us, often in disjointed fragments. The Swedish geneticist Svante Pääbo won a Nobel prize for jigsawing the genome of Neanderthal and Denisovan humans, otherwise known only from fossils; in the Denisovan case only three teeth and five bone fragments. Pääbo's work incidentally shows that Europeans, but not sub-Saharan Africans, are descended from rare cases of interbreeding with Neanderthals. Also, some modern humans, especially Melanesians, can be traced back to interbreeding events with Denisovans. The field of 'Ancient DNA' research is now flourishing. The woolly mammoth genome is almost completely known, and there are serious hopes of reviving the species. Other possible 'resurrections' might include the dodo, passenger pigeon, great auk, and thylacine (Tasmanian wolf). Unfortunately, sufficient DNA doesn't last more than a few thousand years at best. In any case, interesting though it is, Ancient DNA is outside the scope of this book.

Fourth, I shall not be dealing with comparisons of DNA sequences in different populations of modern humans and the light that they throw on history, including the waves of human migration that have swept over Earth's land surface. Tantalisingly, these genetic studies overlap with comparisons between languages. For example, the distribution of both genes and words across the Micronesian islands of the Western Pacific islands shows a mathematically lawful relationship between inter-island distance and word-resemblance. We can picture outrigger canoes scudding across the open Pacific, laden with both genes and words! But that would be a chapter in another book. Might it be called *The Selfish Meme*?

The present book's title should not be taken to mean that existing science is ready to translate DNA sequences into descriptions of ancient environments. Nobody can do that, and it's not clear that SOF will ever do so. This book is about reading the animal itself, its body and behaviour – the 'phenotype'. It remains true that the descriptive messages from the past are transmitted by DNA. But for the moment we read them indirectly via phenotypes. The easiest, if not the only, way to translate a human genome into a working body is to feed it into a very special interpreting device called a woman.

The Species as Sculpture; the Species as Averaging Computer

Sir D'Arcy Thompson (1860–1948), that immensely learned zoologist, classicist, and mathematician, made a remark that seems trite, even tautological, but it actually provokes thought. 'Everything is the way it is because it got that way.' The solar system is the way it is because the laws of physics turned a cloud of gas and dust into a spinning disc, which then condensed to form the sun, plus orbiting bodies rotating in the same plane as each other and in the same direction, marking the plane of the original disc. The moon is the way it is because a titanic bombardment of Earth 4.5 billion years ago hived off into orbit a great quantity of matter, which then was pulled and kneaded by gravity into a sphere. The moon's initial rotation later slowed, in a phenomenon called 'tidal locking', such that we only ever see one face of it. More minor bombardments disfigured the moon's surface with craters. Earth would be pockmarked in the same way but for erosive and tectonic obliteration. A sculpture is the way it is because a block of Carrara marble received the loving attention of Michelangelo.

Why are our bodies the way they are? Partly, like the moon, we bear the scars of foreign insults – bullet wounds, souvenirs of the duellist's sabre or the surgeon's knife, even actual craters from smallpox or chickenpox. But these are superficial details. A body mostly got that way through the processes of embryology and growth. These were, in turn, directed by the DNA in its cells. And how did the DNA get to be the way it is? Here we come to the point. The genome of every individual is a sample of the gene pool of the species. The gene pool got to be the way it is over many generations, partly through random drift, but more pertinently through a process of non-random sculpture. The sculptor is natural selection, carving and whittling the gene pool until it – and the bodies that are its outward and visible manifestation – is the way it is.

Why do I say it's the species gene pool that is sculpted rather than the individual's genome? Because, unlike Michelangelo's marble, the genome of an individual doesn't change. The individual genome is not the entity that the sculptor carves. Once fertilisation has taken place, the genome remains fixed, from zygote right through embryonic development, to childhood, adulthood, old age. It is the gene pool of the species, not the genome of the individual, that changes under the Darwinian chisel. The change deserves to be called sculpting to the extent that the typical animal form that results is an improvement. Improvement doesn't have to mean more beautiful like a Rodin or a Praxiteles (though it often is). It means only getting better at surviving and reproducing. Some individuals survive to reproduce. Others die young. Some individuals have lots of mates. Others have none. Some have no children. Others a swarming, healthy brood. Sexual recombination sees to it that the gene pool is stirred and shaken. Mutation sees to it that new genetic variants are fed into the mingling pool. Natural selection and sexual selection see to it that, as generation succeeds generation, the shape of the average genome of the species changes in constructive directions.

Unless we are population geneticists, we don't see the shifting of the sculpted gene pool directly. Instead, we observe changes in the average bodily form and behaviour of members of the species. Every individual is built by the cooperative enterprise of a sample of genes taken from the current pool. The gene pool of a species is the ever-changing marble upon which the chisels, the fine, sharp, exquisitely delicate, deeply probing chisels of natural selection, go to work.

A geologist looks at a mountain or valley and 'reads' it, reconstructs its history from the remote past through to recent times. The natural sculpting of the mountain or valley might begin with a volcano, or tectonic subduction and upthrust. The chisels of wind and rain, rivers and glaciers then take over. When a biologist looks at fossil history, she sees not genes but things that eyes are equipped to see: progressive changes in average phenotype. But the entity being carved by natural selection is the species gene pool.

The existence of sexual reproduction confers on The Species a very special status not shared by other units in the taxonomic hierarchy – genus, family, order, class, etc. Why? Because sexual recombining of genes – shuffling the pack (American deck) – takes place only within the species. That is the very definition of 'species'. And it leads me to the second metaphor in the title of this section: the species as averaging computer.

The genetic book of the dead is a written description of the world of no particular ancestral individual more than another. It is a description of the environments that sculpted the whole gene pool. Any individual whom we examine today is a sample from the shuffled pack, the shaken and stirred gene pool. And the gene pool in every generation was the result of a statistical process averaged over all those individual successes and failures within the species. The species is an averaging computer. The gene pool is the database upon which it works.

2

'Paintings' and 'Statues'

When, like that Mojave Desert lizard, an animal has its ancestral home painted on its back, our eyes give us an instant and effortless readout of the worlds of its forebears, and the hazards that they survived. Here's another highly camouflaged lizard. Can you see it on its

background of tree bark? You can, because the photograph was taken in a strong light from close range. You are like a predator who has had the good fortune to stumble upon a victim under ideal seeing conditions. It is such close encounters that exerted the selection pressure to put the finishing touches to the camouflage's perfection. But how did the evolution of camouflage get its start? Wandering predators, idly scanning out of the corner of their eye, or hunting when the light was poor, supplied the selection pressures that began the process of evolution

towards tree bark mimicry, back when the incipient resemblance was only slight. The intermediate stages of camouflage perfection would have relied upon intermediate seeing conditions. There's a continuous gradient of available conditions, from 'seen at a distance, in a poor light, out of the corner of the eye, or when not paying attention' all the way up to 'close-up, good light, full-frontal'. The lizard of today has a detailed, highly accurate 'painting' of tree bark on its back, painted by genes that survived in the gene pool because they produced increasingly accurate pictures.

We have only to glance at this frog to 'read' the environment of its ancestors as being rich in grey lichen. Or, in another of Chapter 1's

formulations, the frog's genes 'bet' on lichen. I intend 'bet' and 'read' in a sense that is close to literal. It requires no sophisticated techniques or apparatus. The zoologist's eyes are sufficient. And the Darwinian reason for this is that the painting is designed to deceive predatory eyes that work in the same kind of way as the zoologist's own eyes. Ancestral frogs survived because they successfully deceived predatory eyes similar to the eyes of the zoologist – or of you, vertebrate reader.

In some cases, it is not prey but predators whose outer surface is painted with the colours and patterning of their ancestral world, the better to creep up on prey unseen. A tiger's genes bet on the tiger being born into a world of light and shade striped by vertical stems. The zoologist examining the body of a snow leopard could bet that its ancestors lived in a mottled world of stones and rocks, perhaps a mountainous region. And its genes place a future bet on the same environment as cover for its offspring.

By the way, the big cat's mammalian prey might find its camouflage more baffling than we do. We apes and Old World monkeys have trichromatic vision, with three colour-sensitive cell types in our retinas, like modern digital cameras. Most mammals are dichromats: they are what we would call red-green colour-blind. This probably means they'd find a tiger or snow leopard even harder to distinguish from its background than we would. Natural selection has 'designed' the stripes of tigers, and the blotches of snow leopards, in such a way as to fool the dichromat eyes of their typical prey. They are pretty good at fooling our trichromat eyes too.

Also in passing, I note how surprising it is that otherwise beautifully camouflaged animals are let down by a dead giveaway – symmetry. The feathers of this owl beautifully imitate tree bark. But the symmetry gives the game away. The camouflage is broken.

I am reduced to suspecting that there must be some deep embryological constraint, making it hard to break away from left-right symmetry. Or does symmetry confer some inscrutable advantage in social encounters? To intimidate rivals, perhaps? Owls can rotate their necks through a far greater angle than we can. Perhaps that mitigates the problem of a symmetrical face. This particular photograph tempts the speculation that natural selection might have favoured the habit of closing one eye because it reduces symmetry. But I suppose that's too much to hope for.

Subtly different from 'paintings' are 'statues'. Here the animal's whole body resembles a discrete object that it is not. A tawny frogmouth or a potoo resembling a broken stump of a tree branch, a stick caterpillar sculpted as a twig, a grasshopper resembling a stone or a clod of dry soil, a caterpillar mimicking a bird dropping, are all examples of animal 'statues'.

The working difference between a 'painting' and a 'statue' is that a painting, but not a statue, ceases to deceive the moment the animal is removed from its natural background. A 'painted' peppered moth removed from the light-coloured bark that it resembles and placed on any other background will instantly be seen and caught by a predator. In this photograph, the background is a soot-blackened tree in an industrial area, which is

perfect for the dark, melanic mutant of the same species of moth that you may have noticed less immediately by its side. On the other hand,

the masquerading Geometrid stick caterpillar photographed by Anil Kumar Verma in India, if placed on any background, would have a good chance of still being mistaken for a stick and overlooked by a predator. That is the mark of a good animal statue.

Although a statue resembles objects in the natural background, it does not depend for its effectiveness on being seen against that background in the way that a 'painting' does. On the contrary, it might be in greater danger. A lone stick insect on a lawn might be overlooked, as a stick that had fallen there. A stick insect surrounded by real sticks might be spotted as the odd one out. When drifting alone, the

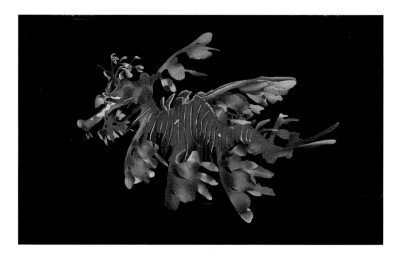

leafy sea dragon's resemblance to a wrack might protect it, at least more so than its seahorse cousin whose shape in no way mimics a seaweed. But would this statue be less safe when nestling in a waving bed of real seaweed? It's a moot question.

Freshwater mussels of the species *Lampsilis cardium* have larvae that grow by feeding on blood, which they suck from the gills of a fish. The mussel has to find a way to put its larvae into the fish. It does it by means of a 'statue', which fools the fish. The mussel has a brood pouch for very young larvae on the edge of its mantle. The brood pouch is an impressive replica of a pair of small fish, complete with false eyes and false, very fish-like, 'swimming' movements. Statues don't move, so the word 'statue' is strictly inappropriate, but never mind, you get the point. Larger fish approach and attempt to catch the dummy fish. What they actually catch – and it does them no good – is a squirt of mussel larvae.

This highly camouflaged snake from Iran has a dummy spider at the tip of its tail. It may look only half convincing in a still picture. But the snake moves its tail in such a way that it looks strikingly like a spider scuttling about. Very realistic indeed, especially when the snake itself is concealed in a burrow with only the tail tip visible. Birds swoop down on the spider. And that is the last thing they do. It is worth reflecting on how remarkable it is that such a trick has evolved by natural selection. What might the intermediate stages have looked like? How did the evolutionary sequence get started? I suppose that, before the tip of the tail looked anything like a spider, simply waggling it about was somewhat attractive to birds, who are drawn to any small moving object.

Both 'paintings' and 'statues' are easy-to-read descriptions of ancestral worlds, the environments in which ancestors survived. The stick caterpillar is a detailed description of ancient twigs. The potoo is a perfect model of long-forgotten stumps. Except that they are not really forgotten. The potoo itself is the memory. Twigs of past ages have carved their own likeness into the masquerading body of that caterpillar. The sands of time have painted their collective self-portrait on the surface of this spider, which you may have trouble spotting.

'Where are the snows of yesteryear?' Natural selection has frozen them in the winter plumage of the willow ptarmigan.

The leaf-tailed gecko recalls to our minds, though not his, the dead leaves among which his ancestors lived. He embodies the Darwinian 'memory' of generations of leaves that fell long before men arrived in Madagascar to see them, probably long before men existed anywhere.

The green katydid (long-horned grasshopper) has no idea that it embodies a genetic memory of green mosses and fronds over which its ancestors walked. But we can read at a glance that this is so. Same with this adorable little Vietnamese mossy frog.

Statues don't always copy inanimate objects like sticks or pebbles, dead leaves, or tree branch stubs. Some mimics pretend to be poisonous or distasteful models, and inconspicuous is precisely what they are not. At first glance you might think this was a wasp and hesitate

to pick it up. It's actually a harmless hoverfly. The eyes give it away. Flies have bigger compound eyes than wasps. This feature is probably written in a deep layer of palimpsest that, for some reason, is hard to over-write. The largest anatomical difference between flies and wasps – two wings rather than four (the feature that gives the fly Order its Latin name, Diptera) – is perhaps also difficult to over-write. But maybe, too, that potential clue is hard to notice. What predator is going to take the time to count wings?

Real wasps, the models for the hoverfly mimicry, are not trying to hide. They're the opposite of camouflaged. Their vividly striped abdomen shouts 'Beware! Don't mess with me!' The hoverfly is shouting the same thing, but it's a lie. It has no sting and would be good to eat if only the predator dared to attack it. It is a statue, not a painting, because its (fake) warning doesn't depend on the background. From our point of view in this book, we can read its stripes as telling us that the ecology of its ancestors contained dangerous yellow-and-black stripy things, and predators that feared them. The fly's stripes are a simulacrum of erstwhile wasp stripes, painted on its abdomen by natural selection. Yellow and black stripes on an insect reliably signify a warning – either true or false – of dire consequences to would-be attackers. The beetle to the right is another, especially vivid example.

If you came face to face with this, peering at you through the undergrowth, would you start back, thinking it was a snake?

It isn't peering and it isn't a snake. It's the chrysalis of a butterfly, *Dynastor darius*, and chrysalises don't peer. As a fine pretence of the front end of a snake, it's well calculated to frighten. Never mind that rational second thoughts could calculate that it's a bit on the small side to be a dangerous snake. There exists a distance – still close enough to be worrying – at which a snake would look that small. Besides, a panicking bird has no time for second thoughts. One startled squawk and it's away. Having more time for reflection, the Darwinian student of the genetic book of the dead will read the caterpillar's ancestral world as inhabited by dangerous snakes. Some caterpillars, whose rear ends pull the same snake trick, even move muscles in such a way that the fake eyes seem to close and open.

Would-be predators can't be expected to know that snakes don't do that.

Eyes are scary in themselves. That's why some moths have eyespots on their wings, which they suddenly expose when surprised by a predator. If you had good reason to fear tigers or other members of the cat family, might you not start back in alarm if suddenly confronted with this, the so-called owl moth of South East Asia?

There exists a distance – a dangerous distance – at which a tiger or a leopard would present a retinal image the same size as a close-up moth. OK, it doesn't look very like any particular member of the cat family to our eyes. But there's plenty of evidence that animals of various species respond to dummies that bear only a crude resemblance to the real thing – scarecrows are a familiar example, and there's lots of experimental evidence as well. Black-headed gulls respond to a model gull head on the end of a stick, as though it were a whole real gull. A shocked withdrawal might be all it takes to save this moth.

I am amused to learn that eyes painted on the rumps of cattle are effective in deterring predation by lions.

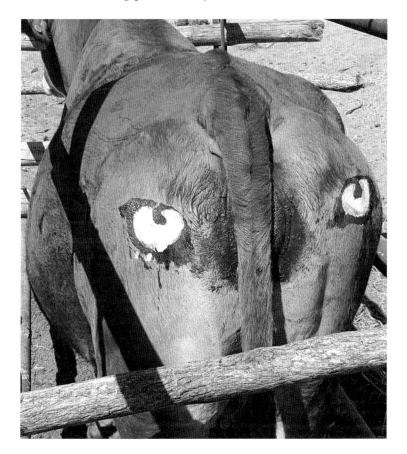

We could call it the Babar effect, after Jean de Brunhoff's lovable and wise King of the Elephants, who won the war against the rhinoceroses by painting scary eyes on elephant rumps.

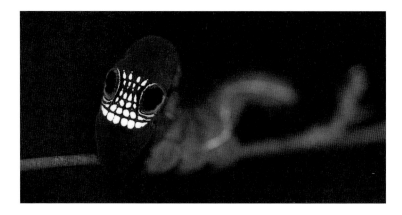

What on Earth is this? A dragon? A nightmare devil horse? It is in fact the caterpillar of an Australian moth, the pink underwing. The spectacular eye and teeth pattern is not visible when the caterpillar is at rest. It is screened by folds of skin. When threatened, the animal pulls back the skin screen to unveil the display, and, well, all I can say is that if I were a would-be predator, I wouldn't hang about.

PHOTO (RIGHT): HUSEIN LATIF

The scariest false face I know? It's a toss-up between the octopus on the left and the vulture on the right. The real eyes of the octopus can just be seen above the inner ends of the 'eyebrows' of the large, prominent false eyes. You can find the real eyes of the Himalayan griffon vulture if you first locate the beak and hence the real head. The false eyes of the octopus presumably deter predators. The vulture seems to use its false face to intimidate other vultures, thereby clearing a path through a crowd around a carcase.

Some butterflies have a false head at the back of the wings. How might this benefit the insect? Five hypotheses have been proposed, of which the consensus favourite is the deflection hypothesis: birds are thought to peck at the less vulnerable false head, sparing the real one. I slightly prefer a sixth idea, that the predator expects the butterfly to take off in the wrong direction. Why do I prefer it? Perhaps because I am committed to the idea that animals survive by predicting the future.

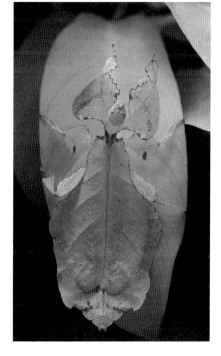

Paintings and statues aimed at fooling predators constitute the nearest approach achieved by any book of the dead to a literal readout, a literal description of ancestral worlds. And the aspect of this that I want to stress is its astounding accuracy and attention to detail. This leaf insect even has fake blemishes. The stick caterpillar (page 19) has fake buds.

I see no reason why the same scrupulous attention to detail should not pervade less literal, less obvious parts of the readout. I believe the same detailed perfection is lurking, waiting to be discovered, in internal organs, in brain-wiring of behaviour, in cellular biochemistry, and other more indirect or deeply buried readings that can be dug out if only we could develop the tools to do so. Why should natural selection escalate its vigilance specifically for the *external appearance* of animals? Internal details, *all* details, are no less vital to survival. They are equally subject to becoming written descriptions of past worlds, albeit written in a less transparent script, harder to decipher than this chapter's superficial paintings and statues. The reason paintings and statues are easier for us to read than internal pages of the genetic book of the dead is not far to seek. They are aimed at eyes, especially predatory eyes. And, as already pointed out, predatory eyes, vertebrate ones at least, work in the same way as our eyes. No wonder it is camouflage and other versions of painting and sculpture that most impress us among all the pages of the book of the dead.

I believe the internally buried descriptions of ancestral worlds will turn out to have the same detailed perfection as the externally seen paintings and statues. Why should they not? The descriptions will just be written less literally, more cryptically, and will require more sophisticated decoding. As with the ear's decoding of Chapter 1's spoken word 'sisters', the paintings and statues of this chapter are effortlessly read pages from books of the dead. But just as the 'sisters' waveform, when presented in the recalcitrant form of binary digits, will eventually yield to analysis, so too will the non-obvious, non-skin-deep details of animals and their genes. The book of the dead will be read, even down to minute details buried deep inside every cell.

This is my central message, and it will bear repeating here. The fine-fingered sculpting of natural selection works not just on the external appearance of an animal such as a stick caterpillar, a tree-climbing lizard, a leaf insect or a tawny frogmouth, where we can appreciate it with the naked eye. The Darwinian sculptor's sharp chisels penetrate every internal cranny and nook of an animal, right down to the sub-microscopic interior of cells and the high-speed chemical wheels that turn therein. Do not be deceived by the extra difficulty of discerning details more deeply buried. There is every reason to suppose that painted lizards or moths, and moulded potoos or caterpillars, are the outward and visible tips of huge, concealed icebergs. Darwin was at his most eloquent in expressing the point.

> It may be said that natural selection is daily and hourly scrutinising, throughout the world, every variation, even the slightest; rejecting that which is bad, preserving and adding up all that is good; silently and insensibly working, whenever and wherever opportunity offers, at the improvement of each organic being in relation to its organic and inorganic conditions of life. We see nothing of these slow changes in progress, until the hand of time has marked the long lapse of ages, and then so imperfect is our view into long past geological ages, that we only see that the forms of life are now different from what they formerly were.

3

In the Depths of the Palimpsest

It's all very well for me to say an animal is a readout of environments from the past, but how far into the past do we go? Every twinge of lower-back pain reminds us that our ancestors only 6 million years ago walked on all fours. Our mammalian spine was built over hundreds of millions of years of horizontal existence when the working body depended on it – depended in the literal sense of hanging from it. The human spine was not 'meant' to stand vertically, and it understandably protests. Our human palimpsest has 'quadruped' boldly written in a firm hand, then over-written all too superficially – and sometimes painfully – with the tracery of a new description – biped. Parvenu, Johnny-come-lately biped.

The skin of Chapter 1's Mojave horned lizard proclaimed to us an ancestral world of sandy, stony desert, but that world was presumably recent. What can we read from the palimpsest about earlier environments? Let's begin by going back a very long way. As with all vertebrates, lizard embryos have gill arches that speak to us of ancestral life in water. As it happens, we have fossils to tell us that the watery scripts of all terrestrial vertebrates, including lizards, date back to Devonian times and then back to life's marine beginning. The poetic point has often been made – I associate it with that salty, larger-than-life intellectual warrior JBS Haldane – that our saline blood plasma is a relic of Palaeozoic seas. In a 1940 essay called 'Man as a Sea Beast', Haldane notes that our plasma is similar in chemical composition to the sea but diluted. He takes this as an indication, not a very strong one in my reluctant opinion ('reluctant' because I like the idea), that Palaeozoic seas were less salty than today's:

> As the sea is always receiving salt from the rivers, and only occasionally depositing it in drying lagoons, it becomes saltier from age to age, and our plasma tells us of a time when it possessed less than half its present salt content.

The phrase 'tells us of a time' resonates congenially with the title of this book. Haldane goes on:

> we pass our first nine months as aquatic animals, suspended in and protected by a salty fluid medium. We begin life as salt-water animals.

Whatever the plausibility of Haldane's inference about changing salinity, what is undeniable is this. All life began in the sea. The lowest level of palimpsest tells a story of water. After some hundreds of millions of years, plants and then a variety of animals took the enterprising step out onto the land. Following Haldane's fancy, we could say they eased the journey by taking their private sea water with them in their blood. Animal groups that independently took this step include scorpions, snails, centipedes and millipedes, spiders, crustaceans such as woodlice and land crabs, insects (who later took a further giant leap into the air) and a range of worms who, however, never stray far from moisture to this day. All these animals have 'dry land' inscribed on top of the deeper marine layers of palimpsest. Of special interest to us as vertebrates, the lobefins, a group of fish represented today by lungfish and coelacanths, crawled out of the sea, perhaps initially only in search of water elsewhere but eventually to take up permanent residence on dry land, in some cases very dry indeed. Intermediate palimpsest scripts tell of juvenile life in water (think tadpole) accompanying adult emergence on land.

That all makes sense. There was a living to be made on land. The sun showers the land with photons, no less than the surface of the sea. Energy was there for the taking. Why wouldn't plants take advantage of it via green solar panels, and then animals take advantage of it via plants? Do not suppose that a mutant individual suddenly found itself fully equipped genetically for life on land. More probably, individuals of an enterprising disposition made the first uncomfortable moves. This was perhaps rewarded by a new source of food. We can imagine them learning to make brief, snatch-and-grab forays out of water. Genetic natural selection would have favoured individuals who were especially good at learning the new ploy. Successive generations would have become better and better at learning it, spending less and less time in the sea.

The general name for learned behaviour becoming genetically incorporated is the Baldwin Effect. Though I won't discuss it further here, I suspect that it's important in the evolution of major innovations generally, perhaps including the first moves towards defying gravity in flight. In the case of the lobe-finned fishes who left the water in the Devonian era around 400 million years ago, there are various theories for how it happened. One that I like was proposed by the American palaeontologist AS Romer. Recurrent drought would have stranded fishes in shrinking pools. Natural selection favoured individuals able to leave a doomed pool and crawl overland to find another one. A point in strong favour of the theory is that there would have been a continuous range of distances separating the pools. At the beginning of the evolutionary progression, a fish could save its life by crawling to a neighbouring pool only a short distance away. Later in evolution, more distant pools could be reached. All evolutionary advances must be gradual. A suffocating fish's ability to exploit air requires physiological modification. Major modification cannot happen in one fell swoop. That would be too improbable. There has to be a gradient of step-by-step small improvement. And a gradient of distances between pools, some near, some a bit further, some far, is exactly what is needed. We shall meet the point again in Chapter 6 and the astonishingly rapid evolution of Cichlid fishes in Lake Victoria. Unfortunately, Romer prefaced his theory by quoting evidence that the Devonian was especially prone to drought. When this evidence was called into question, Romer's whole theory suffered in appreciation. Unnecessarily so.

In whatever way the move to the land happened, profound re-design became necessary. Water really is a very different environment from airy land. For animals, the move out of water was accompanied by radical changes in anatomy and physiology. Watery scripts at the base of the palimpsest had to be comprehensively over-written. It is the more surprising that a large number of animal groups later went into reverse, throwing their hard-won retooling to the winds as they trooped back into the water. Among invertebrates, the list includes pond snails, diving bell spiders, and water beetles. The water that they re-invaded is fresh water, not sea. But some vertebrate returnees, notably whales (including dolphins), sea cows, sea snakes, and turtles, went right back into the salted marine world that their ancestors had taken such trouble to leave.

Seals, sea lions, walruses, and their kin, also Galapagos marine iguanas, only partially returned to the sea, to feed. They still spend much time on land, and breed on land. So do penguins, whose streamlined athleticism in the sea is bought at the cost of risible maladroitness on land. You cannot be a master of all trades. Sea turtles laboriously haul themselves out on land to lay eggs. Otherwise, they

Sea turtle

totally recommitted to the sea. As soon as baby turtles hatch in the sand, they lose no time in racing down the beach to the sea. Lots of other land vertebrates moved part-time into fresh water, including snakes, crocodiles, hippos, otters, shrews, tenrecs, rodents such as water voles and beavers, desmans (a kind of mole), yapoks (water opossums), and platypuses. These still spend a good deal of time on land, taking to the water mainly to feed.

You might think that returnees to water would unmask the lower layers of palimpsest and rediscover the designs that served their ancestors so well. Why don't whales, why don't dugongs, have gills? Their embryos, like the embryos of all mammals, even have the makings of gills. It would seem the most natural thing in the world to dust off the old script and press it into service again. That doesn't happen. It's almost as though, having gone to such trouble to evolve lungs, they were reluctant to abandon them, even if, as you might think, gills would serve them better. Given gills, they wouldn't have to keep coming to the surface to breathe. But rather than revive the gill, what they did was stick loyally to the lung, even at the cost of profound modifications to the whole air-breathing system, to accommodate the return to water.

They changed their physiology in extreme ways such that they can stay under water for over an hour in some cases. When whales do come to the surface, they can exchange a huge volume of air very quickly in one roaring gulp before submerging again. It's tempting to toy with the idea of a general rule stating that old scripts from lower down the palimpsest cannot be revived. But I can't see why this should in general be true. There has to be a more telling reason. I suspect that, having committed their embryological mechanics to air-breathing lungs, the repurposing of gills would be a more radical embryological upheaval, more difficult to achieve than rewriting superficial scripts to modify the air-breathing equipment.

Sea snakes don't have gills, but they obtain oxygen from water through an exceptionally rich blood supply in the head. Again, they

Steller's sea cow

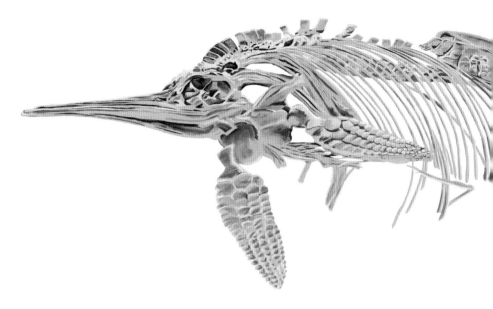

went for a new solution to the problem, rather than revive the old one. Some turtles obtain a certain amount of oxygen from water via the cloaca (waste disposal plus genital opening), but they still have to come to the surface to breathe air into their lungs.

Never parted from the buoyant support of water, whales are freed to evolve in massively (indeed so) different directions from their terrestrial ancestors. The blue whale is probably the largest animal that ever lived. Steller's sea cows (see previous page), extinct relatives of dugongs and manatees, reached lengths of 11 metres and masses of 10 tonnes, larger than minke whales. They were hunted to extinction in the eighteenth century, soon after Steller first saw them. Like whales, sea cows breathe air, having failed to rediscover anything equivalent to the gills of their earlier ancestors. For reasons just discussed, that word 'failed' may be ill-advised.

Ichthyosaurs were reptilian contemporaries of the dinosaurs, with fins and streamlined bodies, and with powerful tails, which were their main engines of propulsion: like dolphins, except that ichthyosaur tails would have moved from side to side rather than up and down. The ancestors of whales and dolphins had already perfected the mammalian galloping gait on land, and the up-and-down motion of dolphin flukes was naturally derived from it. Dolphins 'gallop'

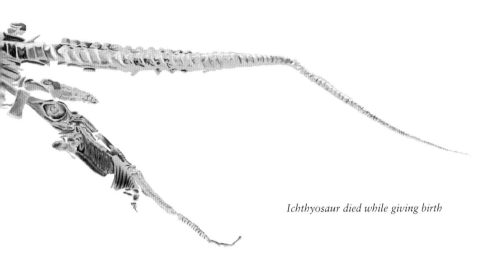

Ichthyosaur died while giving birth

through the water, unlike ichthyosaurs, who would have swum more like fish. Otherwise, ichthyosaurs looked like dolphins and they probably lived pretty much like dolphins. Did they leap exuberantly into the air – wonderful thought – wagging their tails like dolphins (but from side to side)? They had big eyes, from which we might guess that they probably didn't rely on sonar as the small-eyed dolphins do. Ichthyosaurs gave birth to live babies in the sea, as we know from a fossil ichthyosaur who unfortunately died during the act of giving birth (see above). Unlike turtles, but like dolphins and sea cows, ichthyosaurs were fully emancipated from their terrestrial heritage. So were plesiosaurs, for there's evidence that they were livebearers too. Given that viviparity has evolved, according to one authoritative estimate, at least 100 times independently in land reptiles, it seems surprising that sea turtles, buoyant in water but painfully heavy on land, still labour up the sands to lay eggs. And that their babies, when they hatch, are obliged to flap their perilous way down to the sea, running a gauntlet of gulls, frigate birds, foxes, and even marauding crabs.

Sea turtles revert to land to lay their eggs, in holes that they dig in a sandy beach. And an arduous exertion it is, for they are woefully ill-equipped to move out of water. Seals, sea lions, otters, and many other mammals whom we'll discuss in a moment, spend part of their time in water and are adapted to swimming rather than walking, which makes them clumsy on land, though less so than sea turtles. As already remarked, the same is true of penguins, who are champions in water but comically awkward on land. Galapagos marine iguanas are proficient swimmers, but they can manage a surprising turn of speed on land too, when fleeing snakes. All these animals show us what the intermediates might have been like, on the way to becoming dedicated mariners like whales, dugongs, plesiosaurs, and ichthyosaurs.

Tortles and turtoises – a tortuous trajectory

Turtles and tortoises are of special interest from the palimpsest point of view, and they deserve special treatment. But first I have to dispel a confusing quirk of the English language. In British common usage, turtles are purely aquatic, tortoises totally terrestrial. Americans call them all turtles, tortoises being those turtles that live on land. In what follows, I'll try to use unambiguous language that won't confuse readers from either of the two nations 'separated by a common language'. I'll sometimes resort to 'chelonians' to refer to the entire group.

Land tortoises, as we shall see, are almost unique in that their palimpsest chronicles a double doubling-back during the long course of their evolution. Their fish ancestors, along with the ancestors of all land vertebrates including us, left the sea in Devonian times, around 400 million years ago. After a period on land they then, like whales and dugongs, like ichthyosaurs and plesiosaurs, returned to the water. They became sea turtles. Finally, uniquely, some aquatic turtles came back to the land and became our modern dry-land (in some cases very dry indeed) tortoises. This is the 'double doubling-back' that I mentioned. But how do we know? How has the uniquely complicated palimpsest of land tortoises been deciphered?

We can draw a family tree of extant chelonians, using all available evidence including molecular genetics. The diagram below is adapted from a paper by Walter Joyce and Jacques Gauthier. Aquatic groups are shown in blue, terrestrial in orange. I've taken the liberty of colouring the 'ancestral' blobs blue when the majority of their descendant groups are blue. Today's land tortoises constitute a single branch, nested among branches consisting of aquatic turtles.

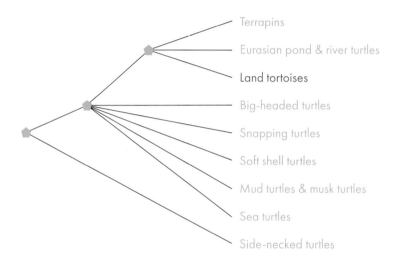

This suggests that modern land tortoises, unlike most land reptiles and mammals, have not stayed on land continuously since their fish ancestors (who were also ours) emerged from the sea. Land tortoises' ancestors were among those who, like whales and dugongs, went back to the water. But, unlike whales and dugongs, they then re-emerged back onto the land. I suppose this means I should reluctantly admit that American terminology has something going for it. As it turns out, what we British call tortoises are just sea turtles who turned turtle and returned to the land. They're terrestrial turtles. No, I can't do it. My upbringing leads me to go on calling them tortoises, but I'll curb my tendency to wince at a phrase like 'desert turtles'. In any case, what is interesting from the point of view of the genetic book of the dead is this: where reversals are concerned, land

Modern land tortoise

tortoises appear to have the most complicated palimpsests of all, with the largest number of almost perverse-seeming reversals.

Moreover, it appears that our modern land tortoises may not be the first of their kind to achieve this remarkable double doubling-back. What looks like an earlier case occurred in the Triassic era. Two genera, *Proganochelys* and *Palaeochersis*, date way back to the first great age of dinosaurs, indeed long before the more spectacular and famous giant dinosaurs of the Jurassic and Cretaceous. It appears that they lived on land. How can we know? This is a good opportunity to return to our 'future scientist' SOF, faced with an unknown animal, and invite her to 'read' its environment from its skeleton. Fossils present the challenge in earnest because we can't watch them living – whether swimming or walking – in their environment.

So, what might SOF say of those enigmatic fossils, *Proganochelys* and *Palaeochersis*? Their feet don't look like swimming flippers. But can we be more scientific about this? Joyce and Gauthier, whom we've already met, used a method that can point the way for anyone who wants to quantitatively decipher the genetic book of the long

Proganochelys

dead. They took seventy-one living species of chelonians whose habitat is known, and made three key measurements of their arm bones, the humerus (upper arm), the ulna (one of the two forearm bones), and the hand, as a percentage of total arm length. They plotted them on triangular graph paper. Triangular plotting makes convenient use of a proof in Euclidean geometry. From any point inside an equilateral triangle, the lengths of perpendiculars dropped to the three sides add up to the same value. This provides a useful technique for displaying three variables when the three are proportions that add up to a fixed number such as one, or percentages that add up to 100. Each coloured point represents one of the seventy-one species. The perpendicular distances of a point from each of the three lines of the big triangle represent the lengths of their three skeletal measurements. And when you colour-code the species according to whether they live in water or on land, something significant leaps off the page. The coloured points elegantly separate out. Blue points represent species living in water, yellow points species living on land. Green points represent genera that spend time in both environments and they, satisfyingly, occupy the region between the blues and yellows.

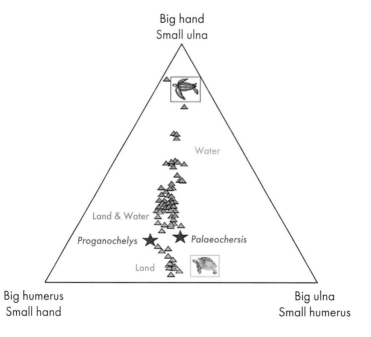

So now, the interesting question is, where do the two ancient fossil species, *Palaeochersis* and *Proganochelys*, fall? They are represented by the two red stars. And there's little doubt about it. The red stars fall among the yellow points, the dry-land species of modern tortoises. They were terrestrial tortoises. The two stars fall fairly close to the green points, so maybe they didn't stray far from water. This kind of method shows one way in which our hypothetical SOF might 'read' the environment of any hitherto unknown animal – and hence read the environment in which its ancestors were naturally selected. No doubt SOF will have more advanced methods at her disposal, but studies such as this one might point the way.

Palaeochersis and *Proganochelys*, then, were landlubbers. But had they stayed on land ever since their (and our) fishy ancestors crawled out of the sea? Or did they, like modern land tortoises, number sea turtles among their forebears? To help decide this, let's look at another fossil. *Odontochelys semitestacea* lived in the Triassic, like *Palaeochersis* and *Proganochelys* but earlier. It was about half a metre long, including a long tail, which modern chelonians lack. The '*Odonto*' in the generic name records the fact that it had teeth, unlike all modern chelonians, who have something more like a bird's beak. And the specific name *semitestacea* testifies to its having only half a shell. It had a 'plastron', the hard shell that protects the belly of all chelonians, but it lacked the domed upper shell. The ribs, however, were flattened like those that support the shell in a normal chelonian.

The fossil was discovered in China and described by a group of scientists led by Li Chun. They believe *Odontochelys*, or something like it, is ancestral to all chelonians and that the turtle shell evolved 'from the bottom up'. They referred to the Joyce and Gauthier paper on forelimb proportions and concluded that *Odontochelys* was aquatic. In case you're wondering what was the use of half a shell, sharks (who have been around since long before any of this story) often attack from below, so the armoured belly might have been anti-shark. If we accept this interpretation, it again suggests that the chelonian shell evolved in water. Against land predators we would not expect that the breastplate should be the first piece of armour to evolve. Quite

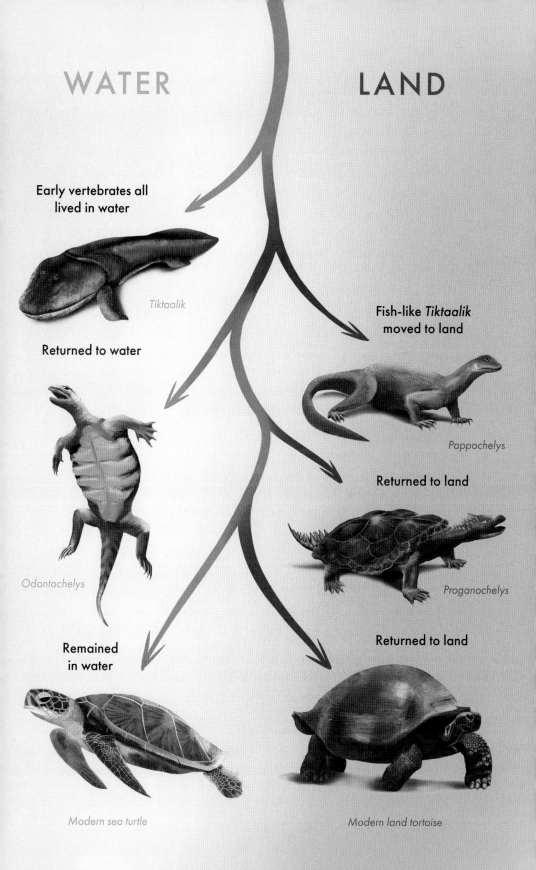

WATER

LAND

Early vertebrates all lived in water

Tiktaalik

Returned to water

Fish-like *Tiktaalik* moved to land

Pappochelys

Odontochelys

Returned to land

Proganochelys

Remained in water

Returned to land

Modern sea turtle

Modern land tortoise

the reverse. *Odontochelys* was probably something like a swimming lizard, a sort of Galapagos marine iguana but armoured with a large ventral breastplate.

Odontochelys

Although it's controversial, the Chinese scientists favour the view that an aquatic turtle like *Odontochelys*, with its half shell, was ancestral to chelonians. Like all reptiles, it would have been descended from terrestrial, lizard-like ancestors, perhaps something like *Pappochelys*. If they are right that the chelonian shell evolved, *Odontochelys*-style, from the bottom up in shark-infested waters, what can we say about *Palaeochersis* and *Proganochelys* out on the land?

It would seem that these represent an earlier emergence from water, an earlier incarnation of doubling-back terrestrial tortoises, to parallel today's behemoths of Galapagos and Aldabra, who evolved from a later generation of aquatic turtles. In any case, the group we know as land tortoises stand as poster child for the very idea of an elaborate palimpsest. Not only did they leave the water for the land, return to water, and then double

Pappochelys

back onto the land again. They may even have done it twice! The doubling-back was achieved first by the likes of *Proganochelys*, and then again, independently, by our modern land tortoises. Maybe some went back to water yet again. It wouldn't surprise me if some freshwater terrapins represent such a triple reversal, but I know of no evidence. Even one doubling-back is remarkable enough.

If this giant Galapagos tortoise could sing a Homeric epic of its ancestors, its DNA-scored Odyssey would range from ancient legends of Devonian fishes, through lizard-like creatures roaming Permian lands, back to the sea with Mesozoic turtles, and finally returning to the land a second time. Now that's what I call a palimpsest!

Giant Galapagos tortoise

Who Sings Loudest

I said in Chapter 1 that the palimpsest chapter would return to the question of the relative balance between recent scripts and ancient ones. It is time to do so. You might conjecture something like the scriptural rule for internal Koranic contradictions: later verses supersede earlier ones. But it's not as simple as that. In the genetic book of the dead, older scripts of the palimpsest can amount to 'constraints on perfection'.

Famous cases of evolutionary bad design, such as the vertebrate retina being installed back to front, or the wasteful detour of the laryngeal nerve (see below), can be blamed on historical constraints of this kind.

'Can you tell me the way to Dublin?'

'Well, I wouldn't start from here.'

The joke is familiar to the point of cliché, but it strikes to the heart of our palimpsest priority question. Unlike an engineer who can go back to the drawing board, evolution always has to 'start from here', however unfavourable a starting point 'here' may be. Imagine what the jet engine would look like if the designer had had to start with a propellor engine on his drawing board, which he then had to modify, step by tinkering step, until it became a jet engine. An engineer starting with the luxury of a clean drawing board would never have designed an eye with the 'photocells' facing backwards, and their output 'wires' being obliged to travel over the surface of the retina and eventually dive through it in a blind spot on their way to the brain. The blind spot is worryingly large, although we don't notice it because the brain, in building its constrained virtual reality model of the world, cunningly fills in a plausible replacement for the missing patch on the visual field. I suppose such guesswork could be dangerous if a hazard happened to fall on the blind spot at a crucial moment. But this piece of bad design is buried deep in embryology. To change it in order to make the end product more sensible would require a major upheaval early in the embryonic development of the nervous system. And the earlier in embryology it is, the more radical and difficult to achieve. Even if such an upheaval could at length be achieved, the intermediate evolutionary stages on the way to the ultimate improvement would probably be fatally inferior to the existing arrangement, which works, after all, pretty well. Mutant individuals who began the long trek to ultimate improvement would be out-competed by rivals who coped adequately with the status quo. Indeed, in the hypothetical case of reforming the retina, they would probably be totally blind.

You can call the backwards retina 'bad design' if you wish. It's a legacy of history, a relic, an older palimpsest script partially over-written. Another example is the tail of humans and other apes, prominent in the embryo, shrunk to the coccyx in the adult. Also faintly traced in the palimpsest is our sparse covering of hair. Once useful for heat insulation, it is now reduced to a relic, still retaining its now almost pointless erectile properties in response to cold or emotion.

The recurrent laryngeal nerve in a mammal or a reptile serves the larynx. But instead of going directly to its destination, it shoots straight past the larynx, on its way down the neck into the chest, where it loops around a major artery and then rushes all the way back up the neck to the larynx. If you think of it as design, this is obviously rotten design. The length of the detour in the giant dinosaur *Brachiosaurus* would have been about 20 metres. In a giraffe it is still impressive, as I witnessed at first hand when, for a Channel Four documentary called *Inside Nature's Giants*, I assisted in the dissection of a giraffe, who had unfortunately died in a zoo. Who knows what inefficiencies or outright errors might have resulted from the transmission delay that such a detour must have imposed. But natural selection is not wantonly silly. It wasn't originally bad design in our fishy ances-tors when the nerve in question went straight to its end organ – not larynx, for fish don't have

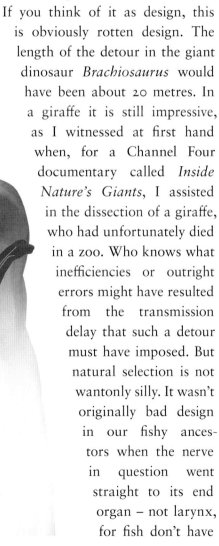

Recurrent laryngeal nerve

a larynx. Fish don't have a neck either. When the neck started to lengthen in their land-dwelling descendants, the marginal cost of each small lengthening of the detour was small compared to what would have been the major cost of radically reforming embryology to re-route the nerve along a 'sensible' path, the other side of the artery. Mutant individuals who began the embryologically radical evolutionary journey towards re-routing the laryngeal nerve would have been out-competed by rival individuals who made do with the working status quo. There's a very similar example in the routing of the tube connecting testis to penis. Instead of taking the most direct route, it loops over the tube connecting kidney to bladder: an apparently pointless detour. Once again, the bad design is a constraint buried deep in embryology and deep in history.

'Buried deep in embryology and deep in history' is another way of saying 'buried deep under layers of younger scripts in the palimpsest'. Far from a 'Koranic' type of rule in which 'Later trumps Earlier', we might be tempted to toy with the reverse, 'Earlier trumps Later'. But that won't do either. The selection pressures that winnowed our recent ancestors are probably still in force today. So, to change the metaphor from a book to a cacophony of voices, the youngest voice, in its youthful vigour, might have something of a built-in advantage. Not an overriding advantage, however. I'd be content with the more cautious claim that the genetic book of the dead is a palimpsest made up of scripts ranging from very old to very young and including all intermediates between. If there are general rules governing relative prominence of old versus young or intermediate, they must wait for later research.

Biologists have long recognised morphological features that lie conservatively in basal layers of the palimpsest. An example is the vertebrate skeleton: the dorsally placed spinal column, with a skull and tail at the two ends, the column made of serially segmented vertebrae through which runs the body's main trunk nerve. Then the four limbs that sprout from it, each consisting of a single, typically long bone (humerus or femur) connected to two parallel bones (radius/ulna, tibia/fibula); then a cluster of smaller bones terminating in five digits. It's always five digits in the embryo, although in the adult some may be reduced or even missing. Horses have lost all but the middle digit, which bears the hoof (a massively enlarged version of our nail). A group of extinct South American herbivores, the Litopterns, included some species, such as *Thoatherium* (left), which independently evolved almost exactly the same hoofed limb as the horse (right). The two limbs have been drawn the same size for ease of comparison, but *Thoatherium* was considerably smaller than a typical horse, about the size of a small antelope. Think of the horse in the picture as a Shetland pony!

Arthropods have a different *Bauplan* (building plan or body plan), although they resemble vertebrates in their segmented pattern of units repeated fore-and-aft in series. Annelid worms such as earthworms, ragworms, and lugworms also have a segmented body plan, and they share with arthropods the ventral position of the main

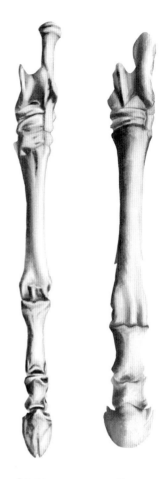

Litoptern   *Horse*

nerve. This difference in position of the body's main nerve has led to the provocative speculation that we vertebrates may be descended from a worm who developed the habit of swimming upside down – a habit that has been rediscovered by brine shrimps today. If this is so, the 'basic' vertebrate *Bauplan* may not be quite as basic as we thought.

Brine shrimp

But, important and even stately as such morphological bauplans are, morphology has become overshadowed by molecular genetics when it comes to reading the lower layers of biological palimpsests in order to reconstruct animal pedigrees. Here's a neat little example. South American trees are inhabited by two genera of tree sloths, the two-toed and the three-toed. There was also a giant ground sloth, which went extinct some ten or twelve thousand years ago, just recently enough to supply molecular biologists with DNA. Since the two tree sloths are so alike, in both anatomy and behaviour, it was natural to suppose that they are closely related, descended from a tree-dwelling ancestor quite recently, and more distantly related to the giant ground sloth. Molecular genetics now shows, however, that the two-toed tree sloth is closer to the giant sloth – all 4 tonnes of it – than it is to the three-toed tree sloth.

Long before modern molecular taxonomy burst onto the scene, morphological evidence aplenty showed us that dolphins are mammals not fish, for all that they look and behave superficially like large fish – mahi-mahi are indeed sometimes called 'dolphinfish' or even 'dolphins'. But although science long knew that dolphins and whales were mammals, no zoologist was prepared for the bombshell released in the late twentieth century by molecular geneticists when they showed, beyond all doubt, that whales sprang from within the artiodactyls, the even-toed, cloven-hoofed ungulates. The closest living cousins of hippos are not pigs, as I was taught as a zoology undergraduate. They are whales. Whales don't have hooves to cleave. Indeed, their land ancestors probably didn't actually have cloven hooves, but broad four-toed feet, as hippos do today. Nevertheless, they are fully paid-up members of the artiodactyls. Not even outliers to the rest of the artiodactyls but buried deep within them, closer cousins to hippos than hippos are to pigs or to other animals who actually have cloven hooves. A staggering revelation that nobody saw coming. Molecular gene sequencing may have other shocks in store for us yet.

Just as a computer disc is littered with fragments of out-of-date documents, animal genomes are littered with genes that must once

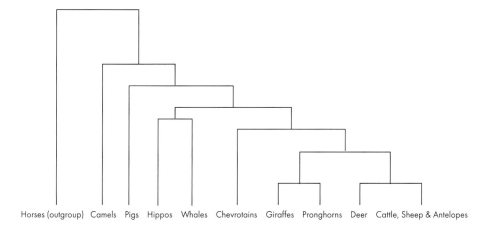

Horses (outgroup) Camels Pigs Hippos Whales Chevrotains Giraffes Pronghorns Deer Cattle, Sheep & Antelopes

Hippos are closer cousins to whales than to any other ungulates

have done useful work but now are never read. They're called pseudo-genes – not a great name, but we're stuck with it. They are also sometimes called 'junk' genes, but they aren't 'junk' in the sense of being meaningless. They are full of meaning. If they were translated, the product would be a real protein. But they are not translated. The most striking example I know concerns the human sense of smell. It is notoriously poor compared with that of coursing hounds, seal-hunting polar bears, truffle-snuffling sows, or indeed the majority of mammals. You'd be right to credit our ancestors with feats of smell discrimination that would amaze us if we could go back and experience them. And the remarkable fact is that the necessary genes, large numbers of them, are still with us. It's just that they are never read, never transcribed, never rendered into protein. They've become side-lined as pseudogenes. Such older scripts of the DNA palimpsest are not only there. They can be read in total clarity. But only by molecular biologists. They are ignored by the natural reading mechanisms of our cells. Our sense of smell is frustratingly poor compared to what it could be if only we could find a way to turn on those ancient genes that still lurk within us. Imagine the high-flown imagery that mutant wine connoisseurs might unleash. 'Black cherry offset by new-mown hay in the attack, with notes of lead pencil in the satisfying finish' would be tame by comparison.

The analogy between genome and computer disc is a more than usually close one. If I invite my computer to list the documents on my hard disc, I see an orderly array of letters, articles, chapters of books, spreadsheets of accounts, music, holiday photos, and so on. But if I were to read the raw data as it is actually laid out on the disc, I would face a phantasmagoria of disjointed fragments. What seems to be a coherent book chapter is made up of here a scrap, there a fragment, dotted around the disc. We think it's coherent only because system software knows where to look for the next fragment. And when I delete a document, I may fondly imagine it has gone. It hasn't. It's still sitting where it was. Why waste valuable computer-time to expunge it? All that happens when you delete a document is that the system software marks its territory on the disc as available to be over-written by other stuff, as and when the space is needed. If the territory is not needed it will not be over-written and the original document, or parts of it, will survive – legible but never actually read – like the smell pseudogenes that we still possess but don't use. This is why, if you want to remove incriminating documents from your computer, you must take special steps to expunge them completely. Routine 'deletion' is not proof against hackers.

Pseudogenes are a lucid message from the past: a significant part of the genetic book of the dead. If she hadn't already deduced it from other cues, SOF would know, from the graveyard of dead genes littering the genome, that our ancestors inhabited a world of smells richer than we can imagine. The DNA tombstones are not only there, the lettering on them is more or less clear and distinct. Incidentally, these molecular tombstones are a huge embarrassment to creationists. Why on earth would a Creator clutter our genome with smell genes that are never used?

This chapter has been mainly concerned with deep layers of the palimpsest, the legacies of more ancient history. In the next four chapters we turn to layers nearer the surface. This amounts to a look at the power of natural selection to override the deep legacies of history. One way to study this is to pick out convergent resemblances between unrelated animals. Another way is 'reverse engineering'. To which we now turn.

4

Reverse Engineering

One of the central messages of this book – that the meticulously detailed perfection we see in the external appearance of animals pervades the whole interior too – obviously rests on an assumption that something approaching perfection is there in the first place. There, and to be expected on Darwinian grounds. It's an assumption that has been criticised and needs defending, which is the purpose of the next three chapters.

The most prominent critics of what they called 'adaptationism' were Richard Lewontin and Stephen Gould, both at Harvard, both distinguished, in their respective fields of genetics and palaeontology. Lewontin defined adaptationism as 'That approach to evolutionary studies, which assumes without further proof that all aspects of the morphology, physiology and behavior of organisms are adaptive optimal solutions to problems.' I suppose I am closer to being an adaptationist than many biologists. But I did devote a chapter of *The Extended Phenotype* to 'Constraints on Perfection'. I distinguished six categories of constraint, of which I'll mention five here.

1. Time lags (the animal is out of date, hasn't yet caught up with a changing environment). Quadrupedal relics in the human skeleton supply one example.

2. Historical constraints that will never be corrected (e.g. recurrent laryngeal nerve, back-to-front retina).
3. Lack of available genetic variation (even if natural selection would favour pigs with wings, the necessary mutations never arose).
4. Constraints of costs and materials (even if pigs could use wings for certain purposes, and even if the necessary mutations were forthcoming, the benefits are outweighed by the cost of growing them).
5. Mistakes due to environmental unpredictability or malevolence (e.g. when a reed warbler feeds a baby cuckoo it is an imperfection from the point of view of the warbler, engineered by natural selection on cuckoos).

If such constraints are allowed for and admitted, I think I could fairly be called an adaptationist. There remains the point, which will occur to many people, that certain 'aspects of the morphology, physiology and behavior of organisms' may be too trivial for natural selection to notice them. They pass under the radar of natural selection. If we are talking about genes as molecular geneticists see them, then it is probably true that most mutations pass unnoticed by natural selection. This is because they are not translated into a changed protein, therefore nothing changes in the organism. They are literally neutral, in the sense of the Japanese geneticist Motoo Kimura, not mutations at all in the functional sense. It's like changing the font in which an instruction is printed, from Times New Roman to Helvetica. The meaning is exactly the same after the mutation as it was before. But Lewontin had sensibly excluded such cases when he specified 'morphology, physiology and behavior'. If a mutation affects the morphology, physiology, or behaviour of an animal, it is not neutral in the trivial 'changing the font' sense.

Nevertheless, some people still have an intuitive feeling that many mutations are probably still negligible, even if they really do affect morphology, physiology, or behaviour. Even if there's a real change visible in the animal's body, mightn't it be too trivial for natural selection to bother about? My father used to try to persuade me that the shapes of leaves, say the difference between oak shape and beech shape, couldn't possibly make any difference. I'm not so sure, and this is where I tend to part company with the sceptics like Lewontin. In 1964, Arthur Cain (my sometime tutor at Oxford) wrote a polemical paper in which he forcefully (some might say too forcefully) argued the case for what he called 'The Perfection of Animals'. On 'trivial' characters, he argued that what seems trivial to us may simply reflect our ignorance. 'An animal is the way it is because it needs to be' was his slogan, and he applied it both to so-called trivial characters and to the opposite – fundamental features like the fact that vertebrates have four limbs and insects have six. I think he was on firmer ground where so-called trivial characters were concerned, for instance in the following memorable passage:

> But perhaps the most remarkable functional interpretation of a 'trivial' character is given by Manton's work on the diplopod [a kind of millipede] *Polyxenus*, in which she has shown that a character formerly described as an 'ornament' (and what could sound more useless?) is almost literally the pivot of the animal's life.

Even in those cases where the character is very close to being genuinely trivial, natural selection may be a more stringent judge than the human eye. What is trivial to our eyes may still be noticed by natural selection when, in Darwin's words, 'the hand of time has marked the long lapse of ages'. JBS Haldane made a relevant hypothetical calculation. He assumed a selection pressure in favour of a new mutation so weak as to seem trivial: for every 1,000 individuals with the mutation who survive, 999 individuals without the mutation will survive. That selection pressure is much too weak to be detected by scientists working in the field. Given Haldane's assumption, how long will it take for such a new mutation to spread through half the population? His answer was a mere 11,739 generations if the gene is dominant, 321,444 generations if it is recessive. In the case of many animals, that number of generations is an eye-blink by geological standards. A relevant point is that, however seemingly trivial a change may be, the mutated gene has very many opportunities to make a difference – via all the thousands of individuals in whose bodies it finds itself over geological time. Moreover, even though a gene may have only one proximal effect, because embryology is complicated, that one primary effect may ramify. As a result, the gene appears to have many seemingly disconnected effects in different parts of the body. These different effects are called pleiotropic, and the phenomenon is pleiotropism. Even if one of a mutation's effects was truly negligible, it's unlikely that all its pleiotropic effects would be.

With all due recognition to the various constraints on perfection, I think a fair working hypothesis is one that, surprisingly, Lewontin himself expressed, admittedly long before his attacks on adaptationism: 'That is the one point, which I think all evolutionists are agreed upon, that it is virtually impossible to do a better job than an organism is doing in its own environment.'

Some biologists prefer to say natural selection produces animals that are just 'good enough' rather than optimal. They borrow from economists the term 'satisficing', a jargon word that they love to namedrop. I'm not a fan. Competition is so fierce, any animal who merely satisficed would soon be out-competed by a rival individual who went one better than satisficing. Now, however, we have to borrow from engineers the important notion of local optima. If we think of a landscape of perfection where improvement is represented by climbing hills, natural selection will tend to trap animals on the top of the nearest relatively low hill, which is separated from a high mountain of perfection by an impassable valley. Going down into the valley is the metaphor for getting temporarily worse before you can get better. There are various ways, known to both biologists and engineers, whereby hill-climbers can escape local optima and make their way to 'broad, sunlit uplands', though not necessarily to the highest peak of all. But I shall leave the topic now.

Engineers assume that a mechanism designed by somebody for a purpose will betray that purpose by its nature. We can then 'reverse engineer' it to discern the purpose that the designer had in mind.

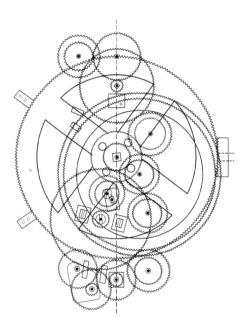

Reverse engineering is the method by which scientific archaeologists reconstructed the purpose of the Antikythera mechanism, a mesh of cogwheels found in a sunken Greek ship dating from about 80 BC. The intricate gearing was exposed by modern techniques such as X-ray tomography. Its original purpose has been reverse engineered as an ancient equivalent of an analogue computer, designed to simulate the movement of heavenly bodies according to the system of epicycles later associated with Ptolemy.

Reverse engineering assumes that the object facing us had a purpose in the mind of a competent designer, a purpose that can be guessed. The reverse engineer sets up a hypothesis as to what a sensible designer might have had in mind, then checks the mechanism to see if it fits the hypothesis. Reverse engineering works well for animal bodies as well as for man-made machines. The fact that the latter were deliberately designed by conscious engineers while the former were designed by unconscious natural selection makes surprisingly little difference: a potential for confusion readily exploited by creationists with their characteristically eager appetite for it. The grace of a tiger and of its prey could not easily, it would seem, be bettered:

> *What immortal hand or eye*
> *Could frame thy fearful symmetry.*

Indeed, animals sometimes seem *too* symmetrically designed, to their own detriment: remember the owl pictured on page 17.

Darwin had a section of *Origin of Species* called 'Organs of extreme perfection and complication'. It's my belief that such organs are the end products of evolutionary arms races. The term 'armament race' was introduced to the evolution literature by the zoologist Hugh Cott in his book on *Animal Coloration* published in 1940, during the Second World War. As a former officer in the regular army during the First World War, he was well placed to notice the analogy with evolutionary arms races. In 1979, John Krebs and I revived the idea of the evolutionary arms race in a presentation to the Royal Society. Whereas an individual predator and its prey run a race in real time, arms races are run in evolutionary time, between lineages of organisms. Each improvement on one side calls forth a counter-improvement on the other. And so the arms race escalates, until called to a halt, perhaps by overwhelming economic costs, just like military arms races.

Antelopes could always outrun lions, and vice versa, but only by counter-productive investment of too much 'capital' in leg muscles at the expense of other calls on investment in, say, milk production. If the language of 'investment' sounds too anthropomorphic, let me translate. Individuals who excel in running speed would be out-competed by slightly slower individuals who divert resources more usefully, from athletic legs into milk. Conversely, individuals who overdo milk production are out-competed by rivals who economise on milk production and put the energy saved into running speed. To quote the economists' hackneyed saw, there's no such thing as a free lunch. Trade-offs are ubiquitous in evolution.

I think arms races are responsible for every biological design impressive enough to, in the words of David Hume's Cleanthes, ravish 'into admiration all men who have ever contemplated them'. Adaptations to ice ages or droughts, adaptations to climate change, are relatively simple, less prone to ravish into admiration because climate is not out to get you. Predators are. So are prey, in the indirect sense that, the more success prey achieve at evading capture, the closer their would-be predators come to starvation. Climate doesn't menacingly change in response to biological evolution. Predators and prey do. So do parasites and hosts. It is the mutual escalation of arms races that drives evolution to Cleanthean heights, such as the feats of mimetic camouflage we met in Chapter 2, or the sinister wiles of cuckoos that will amaze us in Chapter 10.

And now for a point that at first sight seems negative. Whereas animals look beautifully designed on the outside, as soon as we cut them open, we seem superficially to get a different impression. An untutored spectator of a mammal dissection might fancy it a mess. Intestines, blood vessels, mesenteries, nerves seem to spill out all over the place. An apparent contrast with the sinewy elegance of, say, a leopard or antelope when seen from outside. On the face of it, this might seem to contradict the conclusion of Chapter 2. The central point stated there was that the perfection typical of the outer layer must pervade every internal detail as well. Now compare your heart with the village pump, which seems neatly and simply fit for purpose. Admittedly, the heart is two pumps in one, serving the lungs on the one hand and the rest of the body on the other. But you could be forgiven for wondering whether a more minimally elegant pump might profitably have been designed.

Each eye sends information to the brain on the opposite side. Muscles on the left side of the body are controlled by the right side of the brain and vice versa. Why? I suppose we are again dealing with ancient scripts long buried in low strata of the palimpsest. Given such deep constraints, natural selection busily tinkers with the upper-level scripts, making good, as far as possible, the inevitable imperfections imposed by deeper levels. The backwards wiring of the vertebrate

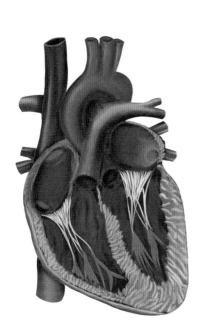

Two pumps

retina is well compensated by *post-hoc* making good. You might think that 'from such warped beginnings nothing debonair can come'. The great German scientist Hermann von Helmholtz is said to have remarked that if an engineer had produced the eye for him, he would have sent it back. Yet after tweaking, 'in post' as movie-makers say, the vertebrate eye can become a fine piece of optical kit.

Why do animals look obviously well designed on the visible outside but apparently less so inside? Does the clue reside in that word 'visible'? In the case of Chapter 2's camouflage, and also ornamental extravaganzas like the peacock's fan, (human) eyes are admiring the external appearance of the animal, and (peahen or predator) eyes are doing the natural selection of external appearance: similar vertebrate eyes in both cases. No wonder external appearance *looks* more perfectly 'designed' than internal details. Internal details are every bit as subject to natural selection, but they don't obviously *look* that way because it is not selection by eyes.

That explanation won't do for the streamlined flair of a sprinting cheetah, or its equally graceful Tommy prey. Those beauties did not evolve for the delectation of eyes but to satisfy the lifesaving requirements of speed. Here it would seem to be the laws of physics that impose what we perceive as elegance: as it is for the aerodynamic grace of a fast jet plane. Aesthetics and functionality converge on the same stylish elegance.

I confess that I find the interior of the body bewilderingly complex. I might even go so heretically far as to dismiss it as a mess. But I am a naive amateur where internal anatomy is concerned. A consultant surgeon whom I have consulted (what else should one do with a consultant?) assures me in no uncertain terms that, to his trained eye, internal anatomy has a beautiful elegance, everything neatly stowed away in its proper place, all shipshape and Bristol fashion. And I suspect that 'trained eye' is exactly the point. In Chapter 1, I contrasted the ear's effortless deciphering of the spoken word 'sisters' with the eye's fumbling impotence to see anything beyond a wavy line on an oscilloscope. My eye sees elegance on the outside. Then when I cut an animal open, my amateur eye contemplates only a mess. The trained sur-

Veins, nerves, arteries, lymphatic system – a whole armful of complexity

geon sees stylish perfection of design, inside as well as out. It is, at least partly, the story of 'sisters' all over again. Yet there is more to be said. Something about embryology.

The sceptic vocally doubts whether it can really matter whether this vein in the arm passes over or under that nerve. Maybe it doesn't in the sense that, if their relationship could be reversed with a magic wand, the person's life might not suffer, and might even improve. But I think it does matter in another sense – the sense that solved the riddle of the laryngeal nerve. Every nerve, blood vessel, ligament, and bone got that way because of processes of embryology during the development of the individual. Exactly which passes over or under what may or may not make a difference to their efficient working, once their final routing is achieved. But the embryological upheaval necessary to effect a change, I conjecture, would raise problems, or costs, sufficient to outweigh other considerations. Especially if the embryological upheaval strikes early. The intricate origami of embryonic tissue-folding and invagination follows a strict sequence, each stage triggering its successor. Who can say what catastrophic downstream consequences might flow from a change in the sequence – the kind of change necessary to re-route a blood vessel, say.

Moreover, perhaps Darwinian forces have worked on human perception to sharpen our appreciation of external appearances as opposed to internal details. At all events, I revert with confidence to the conclusion of Chapter 2. It is entirely unreasonable to suppose that the chisels of natural selection, so delicately adept at perfecting external and visible appearance, should suddenly stop at the animal's skin rather than working their artistry inside. The same standards of perfection must pervade the interior of living bodies, even if less obviously to our eyes. To dissect the non-obvious and make it plain will be the business of future zoological reverse engineers, and it is to them that I appeal.

Ideally, reverse engineering is a systematic scientific project, perhaps involving mathematical models in the sense discussed in Chapter 1. More usually, at present at least, it involves intuitive plausibility arguments. If the object in question has a lens in front of a dark chamber, focusing a sharp image on a matrix of light-sensitive units at the back of the chamber, any person living after the invention of the camera can instantly divine the purpose for which it evolved. But there will be numerous details that will matter and will require sophisticated techniques of reverse engineering, including mathematical analysis. In this chapter our reverse engineering is mostly of the intuitive, common sense kind, like the example of the eye and the camera.

Reverse engineering is supplemented by comparison across species. If SOF is confronted with a hitherto unknown animal, she can read it both by pure reverse engineering ('a device designed by an engineer to do such-and-such would probably look rather like this') and also by comparison with known species ('this organ looks like an organ in so-and-so species that we already know, and it probably is used for the same purpose').

An indirect version of reverse engineering can be used to infer aspects of an animal that cannot be seen, for example when all we have is fossils. We have no fossil evidence about the heart of a dinosaur. But fossils tell us that some sauropods such as *Brontosaurus* and the even larger *Sauroposeidon* had extraordinarily long necks. The CGI artists of *Jurassic Park* beautifully illustrated the dominant view that they reached up to browse tall trees. Like giraffes, only more so. Now the engineer steps in and invokes simple laws of physics to dictate that the heart would have had to generate very high pressure in order to push blood to the height of the animal's brain when plucking leaves from a high tree. You can't suck water through a straw that's more than 10.3 metres tall, even if your sucking is powerful enough to generate a perfect vacuum in the straw. *Sauroposeidon*'s head probably overtopped its heart by about that much, which gives an idea of the pressure that the heart would have had to generate to push blood up to the head. Without ever seeing a fossilised sauropod heart, the engineer infers that it must have generated especially high pressure. Either that or that they didn't browse trees at all.

The difficulty of pumping blood to a head so high made me think of the alleged 'second brain' in the pelvis of some large dinosaurs.

It would have been about on a level with the heart, impressively lower than the head, and requiring much less blood pressure to reach it. Alas, the pelvic brain is probably a nineteenth-century myth. But I never miss an excuse to quote Bert Leston Taylor's delightfully witty poem on the subject.

Behold the mighty dinosaur,
Famous in prehistoric lore,
Not only for his power and strength
But for his intellectual length.
You will observe by these remains
The creature had two sets of brains –
One in his head (the usual place),
The other at his spinal base,
Thus he could reason A priori
As well as A posteriori.
No problem bothered him a bit
He made both head and tail of it.
So wise was he, so wise and solemn,
Each thought filled just a spinal column.
If one brain found the pressure strong
It passed a few ideas along.
If something slipped his forward mind
'Twas rescued by the one behind.
And if in error he was caught
He had a saving afterthought.
As he thought twice before he spoke
He had no judgment to revoke.
Thus he could think without congestion
Upon both sides of every question.
Oh, gaze upon this model beast,
Defunct ten million years at least.

Sauropod blood pressure being hidden behind a 66-million-year wall, we must make do with the next big thing, the giraffe. Though not in the same league as a giant dinosaur, the giraffe's head is quite lofty enough to require an abnormally high blood pressure, out of the ordinary for a mammal. And the following graph bears out the expectation.

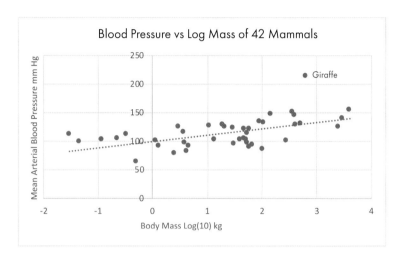

I have plotted mean arterial blood pressure against the logarithm of body mass for a range of mammals from mouse to elephant. It's best to use logarithms for the weights – otherwise it would be hard to fit mouse and elephant on the same page, with intermediate animals conveniently spread out between. The dotted line is the straight line that best fits the data. The line slopes upwards – larger animals tend to have higher blood pressure. Most species are pretty close to the line, meaning that their blood pressure is close to typical for their weight. But the big exception is the giraffe, which is far above the line. Its blood pressure is way higher than it 'should be' for an animal of its size. Surprisingly, other evidence shows that the giraffe heart is not especially large. It seems to be prevented from enlarging in evolution by the need to share the body cavity with large herbivorous guts. It achieves the extra-high blood pressure in a different way, by a greater density of heart muscle cells, an improvement that probably imposes costs of its own. Without ever seeing a *Brontosaurus* heart, we can predict that it too would have stood way above the line in the equivalent graph for reptiles.

The teeth of a hitherto unknown animal speak volumes, and this is fortunate because teeth, being necessarily hard enough to crunch food, are also hard enough to out-last anything else in the fossil record. Some impor-tant extinct species are known only from teeth. In the rest of this chapter, we shall use teeth and other biological food-processing devices as our example of choice. Look at this ancient skull. The first thing you notice is the scary canine teeth. You might reverse engineer these as being good for either fighting rivals or stabbing prey to death and holding onto them. Seeking further evidence, you might then look at the other teeth near the back of the jaw, the molars. They don't mesh surface-to-surface in the way that ours or a horse's do, but shear past each other like scissors as the jaws close. They seem designed to slice rather than to mill. This says 'carnivore'. Well, obvi-ously. But it's only obvious because we are rather good at intuitive reverse engineering, and because we have living large carnivores like lions and tigers for comparison. It does no harm to make the reason-ing explicit.

Sabretooth

Animals, perhaps because they are themselves made of meat, find meat relatively easy to digest, and carnivore intestines tend to be appropriately short. If SOF were handed an unknown animal, very long intestines would signal 'herbivore' to her. I'll return to this. Meat, moreover, demands relatively little pre-processing with teeth before digestion. Cutting off substantial chunks to be swallowed whole is sufficient. Plants may be easier to catch than animals – they don't run away – but they make up for it by being harder to process once you've caught them. Plant cells are different from animal cells. They have thick walls toughened by cellulose and silica. For this and other reasons, herbivores need to grind their food into tiny pieces before it is ready to pass into the gut for further breaking up chemically into even smaller pieces. Herbivore teeth are millstones which, like the mills of God, grind slowly and they grind exceeding small. Carnivore teeth don't resemble millstones and they don't grind. They cut, shearing through fibrous tissues.

Looking at the back teeth of the above skull, then, we confirm our initial diagnosis from the dagger-like canines, and convincingly reverse-engineer our scary specimen as telling a tale of ancestral carnivores. Moving to the rest of the skull, we note that the articulation of the lower jaw allows only up-and-down movement suitable for scissoring food, not side-to-side movement such as would be needed for milling. Up and down is putting it mildly: the sheer size of the gape is formidable. As you'll have guessed, this is the skull of a sabretooth cat, often called sabretooth tiger, although it could just as well be called sabretooth lion. It was a big cat, *Smilodon*, not closer to any particular modern big cat than to any other. Contemporaneous with *Smilodon*, there were true lions in America, now extinct, bigger than *Smilodon*, bigger than African lions.

How did *Smilodon* use those formidable fangs? It's notable that among modern carnivores, the cat family (Felidae) runs to long canine teeth more than the dog family (Canidae), despite the name 'canine' for the teeth. A plausible reason is as follows. Canids are mostly pursuit-hunters. They run their prey down to exhaustion. When they finally catch up with it, the poor spent creature is in no state to escape. Killing it is not a problem. Just start eating! Felids, on the other hand, tend to be stalkers and ambushers. Their prey, when they first pounce upon it, is fresh and in a strong position to escape. Either a swift killing stab or an inescapable grip is desirable, and long penetrating canines answer both needs. Among living cats, the clouded leopard sports the nearest approach to the sabres of *Smilodon*. Clouded leopards spend much of their time in trees and drop on their prey. Long,

sharp daggers would be
especially suited to sub-
duing an animal taken
by surprise from above,
not 'heated in the chase'
and in full possession
of its powers.

Clouded leopard

Turning to other
parts of the skull of
Smilodon, we notice that
the eye sockets point forward,
indicating binocular vision, useful for pouncing on prey and no
good for seeing danger creeping up from
behind. Sabretooths had no need to watch their back. Herbivorous
animals, whose ancestors became ancestors by virtue of noticing
would-be killers, tend to have lookout eyes pointing sideways, giving
almost 360° vision, calculated to spot a predator stalking from any
direction.

So now, suppose you are presented with the skull below. It's obviously very different. The eyes look sideways, as if scanning all around
for danger while not being especially concerned with what is ahead.
Probably an animal with a need to fear predation, then. The incisor
teeth at the front look well suited to cropping grass. Most noticeable are the back teeth. They are broad grinders
rather than sharp slicers, and they
meet their opposite numbers in
a precise fit when the jaws
close. Their whole shape
with its articulation is
well suited to grind-
ing plant food into very
small pieces, again con-
firming the suspicion that
this animal's genes survived

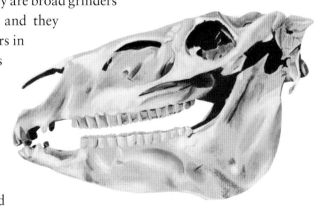

Pliohippus

in a world of grass or other plant food. And the lower jaw, unlike that of *Smilodon*, moves sideways as well as up and down, a good milling action. This fossil is *Pliohippus*, an extinct horse that lived in the Pliocene, probably in mortal fear of *Smilodon*.

The contrast between the skulls of the carnivorous sabretooth and the herbivorous horse is stark and clear. There was an animal called *Tiarajudens*, one of those we used to call a mammal-like reptile (nowadays we'd call it an early mammal), which flourished perhaps 280 million years ago, before the great age of dinosaurs. It had impressive sabretooth canines, much like *Smilodon*, which indicate a carnivorous diet similar to that of the formidable cat. But the back teeth suggest that, along with other animals to whom it was related, it was in fact a herbivore. So, we have a mismatch. Why would a creature with grinding back teeth have canine teeth like *Smilodon*? Perhaps *Tiarajudens* was a herbivore equipped with daggers for defence against predators. Or perhaps, like modern walruses, for fighting against rivals of its own species, as elephants use their gigantic tusks (elephant tusks are enlarged incisor teeth, not canines as in walruses).

Walrus

Walruses have been seen using their (upper canine) tusks to lever themselves out of the water and to make holes in the ice. Anyway,

Tiarajudens stands as a cautionary warning against over-hasty reverse engineering, looking at only one thing, in this case the canine teeth.

Some mammals such as shrews and small bats eat insects. Dolphins eat fish. Though technically carnivorous, the dental demands of these diets are different. Insectivorous teeth are neither grinders nor cutters but piercers. They tend to have sharp points, well suited to piercing the external skeletons of insects. If SOF's unknown specimen sported piercing teeth like those of this hedgehog, she'd suspect that its ancestors survived on a diet of insects and other arthropods. And

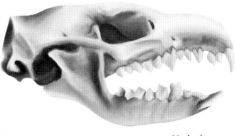

Hedgehog

that is correct, but they like earthworms too. Ants and termites are a special case (see below).

And now here's the skull of a dolphin (top), and a gavial (bottom), to show typical fish-eating teeth and jaws. These two fish-eaters, a mammal and a crocodilian, have independently evolved pretty much the same dentition and jaw shape, an example of convergent evolution (which is

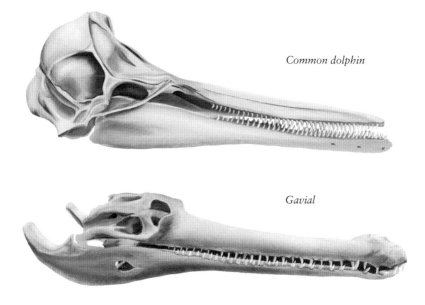

Common dolphin

Gavial

the topic of Chapter 5). What's the reverse-engineering explanation for this convergent resemblance? Fish-eaters, unlike, say, lions, are usually much larger than their prey. They don't need to grind or cut or pierce their prey. Their prey is small enough to swallow whole. Long rows of small, pointed teeth are well equipped to grasp a slippery, soft fish and prevent it from escaping. And the slender jaws can snap shut on the fish without expelling a rush of water that might propel it out of harm's way.

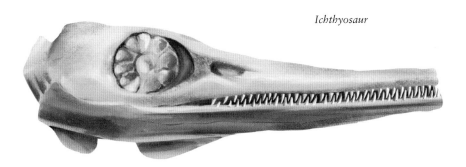

Ichthyosaur

If you were lucky enough to stumble upon a fossil like the above, you could apply the lesson of the previous paragraph: fish-eater. It's an ichthyosaur such as we met in Chapter 3, a contemporary and relative of dinosaurs, member of a large group that went extinct somewhat earlier than the last of the dinosaurs. Both reverse engineering, and comparison with the dolphin and gavial pictures, speak to us loud and clear: its ancestors ate fish.

Killer whales (*Orca*) and sperm whales can be thought of as giant dolphins. They too eat prey smaller than themselves, and they too have long rows of dolphin-like teeth but hugely enlarged. Sperm whales have them only in the lower jaw (very occasionally in the upper jaw, and we may take this as a vestigial relic). Killer whales have them in both jaws. All other large whales, the so-called baleen whales, are filter feeders, sieving krill (crustaceans). They have no teeth at all (though, revealingly, their embryos have them and never use them). Their huge baleen filters are made of keratin, like hooves, fingernails, and rhinoceros horn. The reverse engineer would have no trouble in diagnosing a baleen whale as a trawler. Actually, they are better than trawlers, for they will target a huge aggregation of krill, and gulp it in with copious quantities of sea water, which is then forced out through the curtain of baleen, trapping the krill.

Ants and termites are colossally numerous. A specialist capable of penetrating an ant nest's formidable defences can hoover up a bonanza of food denied to an ordinary insectivore like a hedgehog. And their dentition is correspondingly specialised. For this purpose, by the way, termites are honorary ants. Mammals who preferentially eat ants and/or termites are all called anteaters. There's a group of three South American mammals whose name in English is 'anteater': the Giant Anteater, the Lesser Anteater, and the Silky Anteater.

Giant Anteater

Tamandua

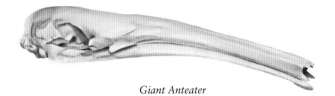

Giant Anteater

Pangolin

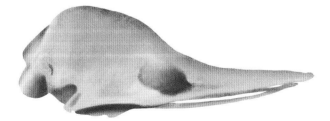

Armadillo

Echidna

The Giant Anteater's scientific name, *Myrmecophaga*, is simply Greek for 'anteater'. You will already have concluded that, since other mammals also specialise in eating ants, 'Anteater' is not a great name for a taxonomic group. I'll use a capital letter for the three South American 'Anteaters' and a lower-case letter for other mammals who eat ants (or termites).

The South American Anteaters push the anteating habit to its extreme. The skulls of two of them, *Tamandua* and the Giant Anteater *Myrmecophaga*, are pictured at the top of the page opposite. Notice the extreme prolongation of the snout and the total absence of teeth. You'd hardly recognise the Giant Anteater's skull as a skull at all. All anteaters show the same features, if to a lesser extent. The pangolin has no teeth and a moderately long snout. Armadillos have a longer snout and rather small teeth. The aardvark or antbear of Africa has back teeth, but no teeth at all along most of its long snout. *Myrmecobius*, the numbat, marsupial anteater of Australia, has a long, pointy head. It has teeth but doesn't use them for eating except in infancy. Adults seem to use them only for gripping and preparing nest material.

Tachyglossus, the spiny anteater or echidna of Australia and New Guinea, is as distant as you can get from all the above while still being a mammal. It's an egg-laying mammal like the platypus, a left-over from the 'mammal-like reptiles' of the ancient supercontinent of Gondwana. But unlike the platypus, with which it shares deep palimpsest features, it does, as its English name suggests, eat ants and termites. And its rather weird-looking skull does indeed have a long, slender snout and no teeth. Let's not get carried away, however. A slightly longer snout is possessed by the related echidna genus, *Zaglossus*, and *Zaglossus* eats almost nothing but earthworms. Evidently, we must be careful before we jump too precipitately to the conclusion that 'long snout' necessarily means anteater. Anteating is not the only habit capable of writing 'long snout' in the palimpsest.

What else might SOF use to diagnose an animal as an anteater? *Myrmecophaga*, the Giant Anteater of South America, whose hugely elongated skull we have already seen, has a giant-sized sticky tongue, which it can protrude to a length of 60 cm, having deployed its formidable claws to break into an ant or termite nest. Huge numbers of the insects stick to the tongue and are drawn in before the tongue shoots out again. Despite its great length, the tongue flicks out and in again at high speed, more than twice per second. Though none can quite match *Myrmecophaga*, creditably long, sticky tongues are also found, convergently evolved, in aardvarks and the unrelated aardwolves, who, unlike other members of the hyaena family, specialise in eating termites. Pangolins, too, have convergently evolved a long sticky tongue. That of the giant pangolin can be 40 cm long and is attached way back near the pelvis instead of to the hyoid bone in the throat, like ours. A pangolin can extend its tongue deep inside an ants' nest, skilfully steering through the labyrinth of tunnels, turning left, turning right, leaving no subterranean avenue unexplored. Tamanduas also have a long sticky tongue but, in this case, their evolution was not independent of *Myrmecophaga*. They surely inherited the long tongue from their shared ancestor, also an anteater. The egg-laying spiny anteater too has a long, sticky tongue, and this time it really is convergent. As is that of the numbat, the marsupial anteater.

There are also physiological resemblances among anteating mammals, notably a low metabolic rate and low body temperature, convergently evolved enough times to impress our hypothetical SOF. However, a low metabolic rate is not exclusively diagnostic of an ant-eating habit. Sloths, befitting their name, also have a low metabolic rate. So do koalas, whom you could regard as a kind of marsupial equivalent of sloths. Both live up trees, eating relatively un-nutritious leaves, and both are slow moving, you might even say lethargic. The convergence doesn't extend to both ends of the alimentary canal, however. Koalas defecate more than a hundred times per day, while sloths hold the record for the other extreme. They defecate about once per week, maybe because they laboriously climb down from the tree in order to do so.

Some of my reverse-engineering conjectures could be wrong. They are only provisional, to illustrate the point that the teeth of an animal, if properly read, will tell a story. In many cases, a story of ancient grassland prairies or leafy forests. Or, if the teeth resemble those of *Smilodon* or the clouded leopard, they speak to us of ambush and stalking. No doubt, if we could read them, every tooth we find could plunge us ever deeper into more specific, detailed stories. Teeth are enamelled archives of ancient history.

Teeth constitute the first food processor in the conveyor belt of digestion. The revealing differences between carnivores and herbivores continue on into the gut. Weight for weight, plants are not so nutritious as meat, so cows, for example, need to graze pretty continuously. Food passes through them like an ever-rolling stream, and they defecate some 40 or 50 kilograms per day. Plant stuff being so different from their own bodies, herbivores need help from chemical specialists to digest it. Those specialist chemists, some of whom were honing their skills perhaps a billion years before animals came on the

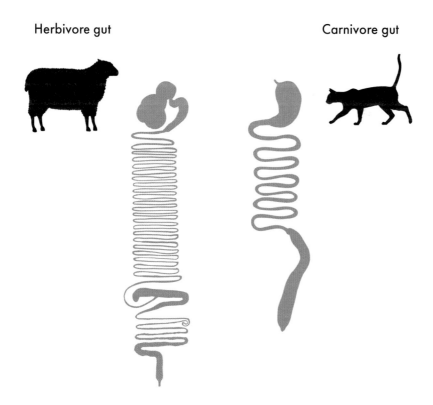

Herbivore gut   Carnivore gut

scene at all, include bacteria, archaea (formerly classified as bacteria but actually far separated from them), fungi, and (what we used to call) protozoa. Ruminants such as cows and antelopes do their fermentation in a different way from horses and rabbits, and at different ends of the gut, but all rely on help from micro-organisms. As already mentioned above, herbivores have longer guts than carnivores, and their guts are complicated by elaborate blind alleys and fermentation chambers, specially fashioned to house symbiotic micro-organisms. Ruminants have the added complication of sending the food back for reprocessing by the teeth for a second time after it's been swallowed – chewing the cud.

There is one bird, the hoatzin of South America, which eats nothing but leaves, the only bird to do so. And – an example of convergent evolution, the process we'll meet in the next chapter – the hoatzin resembles ruminant mammals in having lots of little gut chambers in which are housed bacteria wielding the necessary chemical expertise to digest leaves. Incidentally, there's a widely believed myth that the hoatzin is unique among birds in retaining ancient claws in the front of the wing, like the Jurassic 'intermediate' fossil *Archaeopteryx*. It's true that hoatzin chicks have these primitive claws, but so do the chicks of many other birds, as David Haig pointed out to me. He went on to suggest that this mythic meme is popular among both biologists and creationists, who respectively want *Archaeopteryx* to be, and not to be, an 'evolutionary intermediate'. No animal exists to be primitive for the sake of it, nor to serve as an evolutionary intermediate. The claws are useful to the chicks, who used them for clambering back into a tree when they fall.

Tiktaalik

By the same token animals don't exist for the sake of 'moving on to the next stage in evolution'. The Devonian fossil *Tiktaalik* is widely touted as a transition between fish and land vertebrates. So it may be, but being transitional is not a way to earn a living. Tiktaalik was a living, breathing, feeding, reproducing creature, which should be reverse-engineered as such – not as a half-way stage on the way to something better.

What of our own teeth and jaws, our own guts, and those of our near relatives? What tales of long-gone ancestral meals do they tell? Comparison of our *Homo sapiens* lineage with extinct hominins such as *Paranthropus (Australopithecus) robustus* and *boisei* shows a marked trend over time towards shrinkage of both jaws and teeth in our *sapiens* lineage. The ribcage of those robust old hominins could accommodate a large vegetarian gut. They were evidently less carnivorous than we are, equipped with large plant-milling teeth, strong grinding jaws, and correspondingly powerful jaw muscles. Even though the muscles themselves have not fossilised, their bony attachments, sometimes culminating in a vertical ('sagittal') crest like a gorilla's to increase their purchase, speak to us eloquently of generations of plant roughage. Our own jaw muscles don't reach so high up the side of our head and we have no bony crest.

The primatologist Richard Wrangham has promoted the intriguing hypothesis that the invention of cooking was the key to human uniqueness and human success. He makes a persuasive case that our reduced jaws, teeth, and guts are ill-suited to either a carnivorous or a herbivorous diet unless a substantial proportion of our food is cooked. Cooking enables us to get energy from foods more quickly and efficiently. For Wrangham it was cooking that led to the dramatic evolutionary enlargement of the human brain, the brain being by far the most energy-hungry of our organs. If he's right, it's a nice example of how a cultural change (the taming of fire) can have evolutionary consequences (the shrinking of jaws and teeth).

Birds have no teeth, nor bony jaws. Surprising as it sounds, they may have lost them to save weight – an important concern in a flying animal – replacing them with light, horny beaks. The word 'mandible' is used for both parts of the beak – the upper mandible and the lower mandible. Beaks can tear but they can't chew. Birds do the equivalent of chewing with the gizzard, a muscular chamber of the gut, often containing hard gastroliths – stones or grains of sand that

1. *Macaw*
2. *Crossbill*
3. *Spoonbill*
4. *Eagle*
5. *Skimmer*
6. *Hummingbird*

the bird swallows to help with the milling process. Ostriches swallow appropriately large stones, up to 10 cm. Being flightless, they don't have to worry so much about weight. Even larger stones found with fossil birds such as the giant moas of New Zealand are identified as gastroliths by their polished surfaces – polished by the grinding action in the gizzard.

Beaks vary greatly, and speak to us eloquently of different ways of procuring food. Their variety has been compared with the set of pliers in a mechanic's toolkit. Pointed beaks delicately select small targets such as single seeds or grubs. Parrot beaks are robust nut-crackers or large seed crushers, and the curved upper mandible with its pointed tip is used as something like a hand. Caged parrots can often be seen climbing on the bars, levering themselves up with the beak as if it were a hand. In the wild they use the same trick in trees. Hummingbird beaks are long tubes for imbibing nectar. Imperious, hooked eagle beaks rip flesh from carcases. Woodpecker beaks hammer like high-powered pneumatic drills, pounding rhythmically into trees in search of larvae. They have specially reinforced skulls to cope with the shock of hammering. Flamingo beaks are upside-down filters for small crustaceans, the bird world's nearest approach to the krill-sieving baleen of whales. Oystercatchers use their long, pointed beaks to chisel into mussels and other shellfish. Curlews use theirs to probe mud for worms and shellfish. Spoonbills have flat paddle-like bills that they sweep from side to side, at the same time using their feet to stir up mud and expose small animals lurking in it. Skimmer beaks are even more specialised. The lower mandible is longer than the upper. The bird flies close to the water with the mouth open and the tip of the lower mandible skimming the surface. When it hits a fish, the beak snaps shut, trapping the fish. Pelicans have a volumi-nous pouch of skin under the beak, which nets fish.

Nestling birds who are fed by their parents don't need beaks to do anything other than gape. Their beaks are grotesquely wide, with brightly coloured linings – advertising surfaces garishly designed to out-compete their siblings for parental largesse. The huge difference from adult beaks of the same species reminds us that juvenile needs

Large ground finch

Medium ground finch

Small tree finch

Green warbler finch

can be very different from adult ones, a principle writ large by caterpillars and butterflies, tadpoles and frogs, and many other examples where larval forms occupy a completely different niche from the adults they become.

Crossbills sport a weird crossover of upper and lower jaw beaks, which is helpful in prying apart the scales of pinecones. Insectivorous birds have differently shaped beaks from seed-eaters. And specialists on seeds of different sizes have correspondingly different beaks, the differences making total sense from a reverse-engineering point of view. The evolution of such differences is the subject of a beautiful and still proceeding long-term study of 'Darwin's Finches' on one of the smaller Galapagos Islands by Peter and Rosemary Grant, and their collaborators.

Galapagos is matched as a Pacific island showcase of Darwinian evolution by the archipelago of Hawaii. Both island chains are volcanic and very young by geological standards. The biology of Hawaii differs in being more contaminated by humans, and by the other invasions for which humans are to blame. The evolutionary divergence of Hawaiian honeycreepers (right) shows a variety of beaks that outdoes even that of the Galapagos finches (left). There are eighteen surviving species (more than twice that number have gone extinct), all apparently descended from a single species of Asian finch, probably looking not unlike a Galapagos finch. The range of bill types that has evolved in such a short time is astonishing.

Laysan finch

Some have retained the seed-eating habits of the ancestor, and still look finch-like with stout, stubby beaks. Others have modified their beaks for nectar-sipping, like African sunbirds rather than like New World hummingbirds. Yet others, with long downward-curving beaks, are probers for insects. Of these, the so-called 'I'iwi' (lower right) has a sharp, stout, stabbing lower mandible, which hammers into bark. Then the long curved upper mandible, which has been held out of the way during the hammering, comes into action to probe insects out of the cracks. The Maui parrotbill uses its powerful callipers to crush twigs and rip off bark in search of insects.

Kakawahie (extinct)

Heron beaks are long fishing spears, stabbing down into the water with sudden precision. The African black heron uses its wings to shade its field of view, which would otherwise be troubled by reflections from the rippling water surface. It dramatically sweeps its black wings across its body, laughably recalling a

'Akiapola'au

'I'iwi

black-cloaked villain in Victorian melodrama. A separate problem for anyone spearfishing from above is refraction at the water surface – the illusion that makes oars look bent. There is some evidence that herons and kingfishers adjust their aim to compensate. The archer fishes of Southeast Asia face the same problem in reverse. They lurk under water and shoot insects sitting on tree branches above the surface, by squirting a sudden jet of water straight at the target. That's remarkable enough in itself. Even more so, they seem to compensate for refraction, like herons but in the other direction.

Reverse engineering, then, is one method by which we can read the body of an animal. Another method is to compare it with other animals, both related and unrelated. We used this method to some extent in this chapter. When the genetic books of unrelated animals spell the same message about their environment and way of life, we call it convergence. Convergent resemblances can be spectacular, as we'll see in the next chapter.

Archer fish

5

Common Problem, Common Solution

This book's main thesis is that every animal is a written description of ancestral worlds. It rests upon the hidden assumption – well, not so very hidden – that natural selection is an immensely powerful force, carving the gene pool into shape, deep down among the smallest details. As we saw in Chapter 2, among the most convincing evidence for the power of natural selection is the perfection of camouflage, the consummate detail with which some animals resemble their (ancestral) environment, or resemble an object in that environment. Equally impressive is the detailed resemblance of an animal to another, unrelated animal, because both have converged on the same way of life. Matt Ridley's *How Innovation Works* documents how our greatest human innovations have been hit upon many times independently by inventors in different countries, working in ignorance of each other's efforts. Just the same is true of evolution by natural selection. This chapter is about convergent evolution as an eloquent witness to the power of natural selection.

Despite appearances, the animal above opposite is not a dog. It is an unrelated marsupial, *Thylacinus*, the Tasmanian wolf (often called Tasmanian tiger, for no better reason than the stripes). In (what hindsight can now see as) a heinous crime against nature, the

Thylacine

Tasmanian government in 1888 put a bounty on thylacine heads. The last one to be seen in the wild was infamously shot in 1930 by someone called Wilf Batty. He must have known it was almost extinct, though he couldn't have known his victim was the last one. I suppose in 1930 people still didn't care about such things, a poignant example of what I have called the shifting moral *Zeitgeist*. A captive specimen called Benjamin survived in Hobart Zoo until 1936. *Thylacinus* is one of the best-known examples of convergence. It looked like a dog because it had the same way of life as a dog. Its skull especially is so like a dog's that it is a favourite trick question in zoology student examinations. Such a favourite, indeed, that in my year at Oxford they gave us a real dog skull as a double bluff, assuming that we'd automatically plump for *Thylacinus*.

Rhinoceros beetle

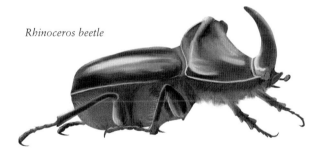

You'd never mistake this for a rhinoceros. But if you watched two rhinoceros beetles fighting, and then two rhinoceroses, you'd realise that convergent resemblances can vault over many orders of magnitude of body size. A fight is a fight is a fight, and a horn is a handy weapon at any size. The same goes for stag beetles and stags, with a somewhat dramatic embellishment. Stag beetles, but not stags, can lift their rivals high in the air on the prongs of their 'antlers'.

Paca   *Chevrotain*

On the left is a paca, a rodent from the rainforests of South and Central America. To its right is a chevrotain or 'mouse deer', an even-toed ungulate that lives in Old World forests. They look like each other convergently because they have similar ways of life. In Africa, the niche is filled by a small ungulate, in South America, by a large rodent.

Armadillos are South American mammals, armoured against predators. When threatened, they roll up into a ball. The picture to the left shows the three-banded armadillo, which rolls up with especially compact elegance. In one of its illustrative quotations, the *Oxford English Dictionary* startlingly records that 'Formerly the armadillo was used in medicine, being swallowed as a pill in its rolled-up state.' Quite a stretch! Until you realise that 'armadillo' in this 1859 quotation referred not to the mammal but to a convergent crustacean, a woodlouse, whose Latin name *Armadillidium* means 'little armadillo'. Armadillo itself is a Spanish word, a diminutive of *armado* or 'armed'. So *Armadillidium* is a diminutive of a diminutive, a double diminutive. The commonality of name speaks to the power of convergent evolution. As befits its vernacular name of 'pill bug', in its rolled-up state you could

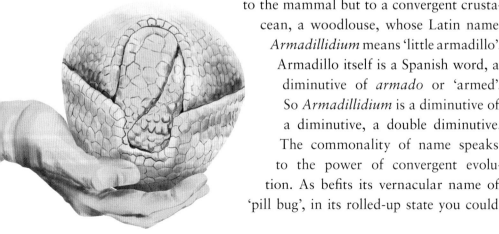

indeed swallow a woodlouse whole, although as to its alleged medicinal value, I shall not comment. The mammalian armadillo and the crustacean *Armadillidium* have converged in their evolution, independently hitting on the same protective habit, albeit at very different sizes, rolling themselves into a ball.

The Latin language has the virtue of condensing into one word what might take three in a language such as English. Latin even has a specialised verb, *glomero*, meaning 'I roll into a ball' (from which we get English words like conglomerate and agglomerate). And *Glomeris* is the scientific name of yet another animal that rolls itself into a ball, and is also called 'pill' in vernacular English. It is not a crustacean but a millipede, the 'pill millipede', a member of the order Glomerida. As if that wasn't enough, two different orders of millipede have independently converged on the roll-up pill body. In addition to the order Glomerida, members of the order Sphaerotheriida (Greek 'spherical beast') look just like *Glomeris* and indeed like *Armadillidium*, except that they are bigger.

Pill woodlouse   *Pill millipede*

Pill woodlouse (above left) and pill millipede (above right) provide what may be my favourite example – in a strong field – of convergent evolution. They are almost indistinguishable when you see them crawling along, or when they roll into a ball. But the one is a crustacean, related to shrimps and crabs, while the other is a myriapod, related to centipedes. To make sure which is which, I have to turn them over. The crustacean has only one pair of legs per segment, making seven pairs in all. The millipede has many more legs, two pairs per segment. These two deeply different 'pill' animals look extremely alike in their surface palimpsest layers because they make their living in the same kind of way and in the same kind of place. Starting from widely separated ancestors they *converged*, in evolutionary time, on very similar end points.

Giant isopod

The deep palimpsest layers show that one is unmistakeably an isopod crustacean, the other a myriapod. Isopods are an important group of crustaceans, and they include members who grow to alarmingly large size on the sea bottom. We shall refer to them again in the next chapter, which goes to town on crustaceans.

Latin isn't the only language to impress with its parsimony. The Malay noun *pengguling* means 'one who rolls up' and from it we get the name pangolin. We met the pangolin in the previous chapter. You might mistake it for a large, animated fir cone. It is not closely related to any other mammals but is out in its own order, Pholidota. That name comes from a Greek word meaning 'covered with scales', and an alternative English name for pangolin is 'scaly anteater'. The scales are made of keratin, like hooves and fingernails. They aren't as hard as the bony armour plates of armadillos.

However, when it comes to glomerising, pangolins perhaps outdo armadillos, pill woodlice, and pill millipedes. According to a report by a biologist on the island of Siberut in Indonesia, a pangolin ran away from him to the top of a steep slope, then formed itself into a ball and rolled down the slope at a speed of about 3 metres per second, twice as fast as a pangolin can run. The witness of this event interpreted the rolling down the hill as a normal response to predation. I reluctantly wonder if it might have been accidental.

There seems to be no doubt as to the effectiveness of rolling up as protection. Lions engage in futile endeavours to penetrate a pangolin's defence. The pangolin's enviable insouciance makes one wonder why other hunted animals don't adopt the same strategy – the tortoise or armadillo strategy – instead of frantically fleeing. I suppose armour is expensive to make, but then so are long, well-muscled, fast-running legs. And it's not a good argument – though possibly true – that if all antelopes, say, were to jettison speed for armour-plated roll-ups, lions on their side of the evolutionary arms race would come up with a counter-strategy. What might be a better argument is that the first individual antelopes to essay rudimentary, and still inadequate, armour would suffer compared with unencumbered rival antelopes disappearing in a cloud of dust.

Lion thwarted by pangolin

Two of the best-known examples of convergent evolution, too familiar to need detailed illustration yet again, are flight and eyes. The laws of physics allow the possibility of using energy to stay aloft for indefinite periods, and the wing has been independently and convergently invented five times: by insects, pterosaurs, birds, bats, and … human technology.

Eyes have been independently evolved many dozens of times, to nine basic designs. The convergent similarity between the camera, the vertebrate eye, and the cephalopod eye has become almost legendary. Here I'll just mention that the most revealing difference – the vertebrate retina but not the mollusc one being wired up backwards – is a difference at a deep palimpsest level. This is another way of saying there's a fundamental difference in their embryology. The vertebrate eye develops mostly as an outgrowth of the brain, while the cephalopod eye develops as an invagination from the outside. That difference lies deep down among the oldest palimpsest layers.

A less familiar example of convergence, compound eyes, have also evolved independently several times. Some bivalve molluscs have a form of compound eye, as do some tube-dwelling annelid worms. These are convergent on each other and on the more highly developed compound eyes of crustaceans, insects, trilobites, and other arthropods. Camera eyes have one lens, which focuses an upside-down image on a retina. The image of a compound eye, if you can call it an image, is the right way up. Think hunting dragonfly, with its pair of large hemispheres, each a cluster of tubes radiating outwards in different directions. Whichever tube sees the target, that's the direction to fly in order to catch it.

A familiar sight throughout both North and South America is the 'turkey vulture'. It looks like a vulture, behaves like a vulture, lives the life of a vulture, feeding on carrion that it finds, like a vulture, with a sense of smell keener than is typical among birds. But it is not a vulture. Or rather, it has converged on vulturehood independently of true vultures. But wait, who is to say that Old World vultures are any more 'true' than New World turkey vultures? Americans might see the priority differently. Let us call both of them vultures, in enthusiastic recognition of convergent evolution and its impressive power to mislead.

We could settle much the same argument about which are the 'true' porcupines. Old World and New World porcupines are both rodents. But within the very large order of rodents, they are not particularly closely related, and they evolved their spiny defences independently. The two pictures show a leopard about to suffer the same punishment from an Old World porcupine as the dog has endured from a New World porcupine.

Dog after approaching New World porcupine

Contrary to legend, no porcupine shoots its quills. But they do have a quick-release mechanism so that a predator injudicious enough to molest a porcupine comes away with a face full of quills. New World quills prolong the agony by means of backward-facing barbs, which make them difficult to remove. This detail is not shared by the otherwise convergent Old World porcupines but it is convergent, at a much smaller scale, on the barbs of bee stings (American stingers).

Leopard approaching Old World porcupine

The sting of a bee, unlike a porcupine quill, is double. There are two barbed blades rubbing against each other with the venom running between them. The two move alternately against each other, sawing their way into the victim. Both are serrated with backward-pointing barbs like those on a New World porcupine quill. The sting is a modified ovipositor, a tube for egg-laying. Porcupine quills are modified hairs. Bees are not the only insects whose ovipositors are serrated. In cicadas (which don't sting), the serrations, and the bee-like alternate sawing action of the two blades, serve to dig the ovipositor (egg-laying tube) into (for example) a tree, where the eggs are laid.

The sting of a bee, derived from the ovipositor and therefore possessed only by females, is a hypodermic syringe for injecting venom. The hypodermic venom injector has evolved convergently in eleven different animal groups by my count (probably more than once independently in some groups): in insects, scorpions, snakes, lizards, spiders, centipedes, stingrays, stonefish, cone shells, and the hind-leg claw of the male duckbilled platypus. The stinging cells, 'cnidoblasts', of jellyfish are miniature harpoons that shoot out on the ends of threads, and inject venom. Among plants, stinging nettles have miniature hypodermic syringes.

The short spikes of hedgehogs are like the long quills of porcupines in being modified hairs. And these too have arisen independently at least three times. There are spiky tenrecs in Madagascar, which look remarkably like hedgehogs although they are not members of the same Order as hedgehogs. They are Afrotheres, related to elephants, aardvarks and dugongs. A third convergence is provided by the spiny anteaters of Australia and New Guinea. Egg-layers, they are as distant from hedgehogs and tenrecs as it is possible to be while still being mammals. They too are covered with spikes, again modified hairs.

We have seen that porcupine quills are a nice example of convergent evolution, independently arisen within the rodents. So-called flying squirrels also arose twice independently in different families of rodents, the true squirrels, and the so-called scaly-tails or anomalures. We know they evolved their gliding habit independently of each other because the closest relatives of both, within the rodents, are not gliders. It's the same way we know New World and Old World porcupines are convergent, again within the large order of rodents.

Not surprisingly, the gliding skill has evolved convergently in a number of vertebrates. The picture shows four mammal examples, including the two rodents just mentioned. The colugo of the Southeast Asian forests is sometimes called the flying lemur, but it isn't a lemur (all true lemurs come from Madagascar, though that's not what makes the colugo a non-lemur) and it doesn't really fly, although it

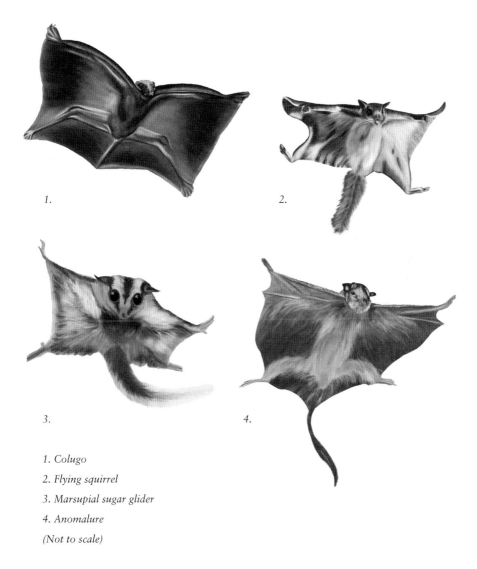

1.

2.

3.

4.

1. *Colugo*

2. *Flying squirrel*

3. *Marsupial sugar glider*

4. *Anomalure*

(Not to scale)

is perhaps a more accomplished glider than the others in the picture. The sugar glider, although it looks extremely like a flying squirrel, is actually a marsupial from Australasia, one of several 'flying phalangers'. Despite the startlingly close resemblance between sugar glider and flying squirrel, we know that one is a marsupial and the other a rodent, because of deeper layers of palimpsest. For example, the female phalanger has a pouch, the squirrel a placenta.

The Australian marsupial fauna provides many other examples of convergent evolution, of which perhaps the most famous is the extinct thylacine or Tasmanian wolf, already mentioned. The picture opposite shows a selection of comparisons between Australian marsupials and their placental equivalents in the rest of the world. These include a pair of anteaters and a pair of 'mice'. The marsupial 'mole' of Australia resembles not only the familiar Eurasian mole but also the 'golden mole' of South Africa. Also very mole-like, among the rodents there are the zokors of Asia.

All these 'moles' independently adopted the same burrowing way of life, all have adapted their hands into powerful spades and all four look pretty alike. So convincing is the convergence that the golden moles were once classified as moles until it was realised that they belong to a radically different branch of (African) mammals, the Afrotheria, together with elephants, aardvarks, and manatees. Eurasian moles, by contrast, are Laurasiatheres, related to hedgehogs, horses, dogs, bats, and whales. Rodent zokors are related to the blind mole rats, who are thoroughly committed to subterranean life and look like moles, but, as you might expect from a rodent, they dig with their teeth rather than their hands. The family tree, overleaf, showing the affinities of four 'moles' is quite surprising.

Impressive as are the convergences of Australian marsupials with a whole variety of placental mammals, we mustn't overlook the exceptions. Kangaroos don't look very like the African antelopes with whom they share a way of life. They easily might have converged. But they didn't. They diverged, mostly because they early committed themselves to a different gait for travelling fast. I suppose there was a time when the ancestors of either could have adopted the hopping gait

Dog   *Thylacine*

European mole   *Marsupial mole*

Mouse   *Marsupial mouse*

Flying squirrel   *Sugar glider*

Tamandua   *Numbat*

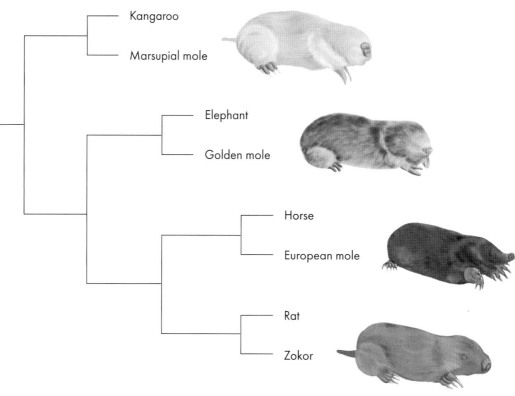

Kangaroo

Marsupial mole

Elephant

Golden mole

Horse

European mole

Rat

Zokor

Independently evolved 'moles'

of a kangaroo or the galloping gait of an antelope. Both gaits are fast and efficient, at least after many generations of evolutionary perfecting. But once an evolutionary lineage starts down a path like hopping or galloping, it is difficult to change. 'Commitment' really is a thing, in evolution. Once a lineage of mammals had advanced some way along the hopping gait path, any mutant that tried to gallop would have been out-competed. Perhaps its front legs were already too short. Conversely, in a lineage that was somewhat committed to galloping, a mutant that tried to hop would clumsily fail. There's no rule that says placental mammals couldn't have taken the kangaroo route. Indeed, there are rodents whose ancestors travelled that path very successfully. A colleague teaching zoology at the University of Nairobi said in a lecture that there were no kangaroos in Africa. This was denied by a student who excitedly claimed to have seen a small one. What he had seen was a springhaas or springhare, a rodent that looks and hops just like a wallaby, complete with foreshortened arms and enlarged, counterbalancing tail.

Springhare

If you could witness an ichthyosaur sporting in Mesozoic waves, you'd be irresistibly reminded of dolphins. A classic case of convergent evolution. On the other hand, your time machine might also present to you a plesiosaur. Far from looking like a dolphin or an ichthyosaur, it doesn't resemble anything else you ever saw. Ichthyosaurs and plesiosaurs are both descended from land reptiles that went back to the sea. But they started out along, and then became 'committed to', alternative paths towards efficient swimming 'gaits'. Ichthyosaurs rediscovered the ancient side-to-side tailbeat of their fish ancestors. They probably passed through a phase resembling the serpentine wavy motion of Galapagos marine iguanas. Plesiosaurs, instead, relied like sea turtles on their limbs, all four of which became huge flippers. Once committed, both ichthyosaurs and plesiosaurs became increasingly dedicated to their respective evolutionary pathways. And ended up looking extremely different.

Convergently evolved animals are not necessarily contemporaries. In North America in the Eocene period there were mole-like subterranean animals, the Epoicotheriids, with mole-like digging hands, not closely related to any living burrowers but belonging to the pangolin family, Pholidota. I'd be surprised if there weren't dinosaur 'moles', but I must confess I don't know of any. There were smallish dinosaurs such as *Oryctodromeus* who dug burrows, but I don't know of any who could be called convergent on moles.

Then there were the so-called 'false sabretooths'. We've already met *Smilodon*, the sabretooth 'tiger', that large, robust and doubtless frightening cat, which went extinct along with most of the American

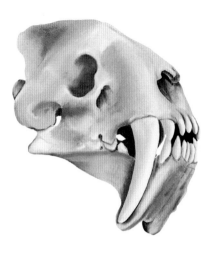
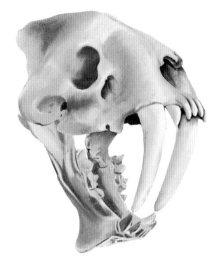

'False' sabretooth – Nimravid

megafauna at the end of the Pleistocene era, only about 10,000 years ago, when man discovered America. What is less well known is that *Smilodon* was not the only member of the order Carnivora to evolve such terrifying fangs. Thirty million years earlier, spanning the Oligocene epoch, lived a group called Nimravids. The Nimravids were not cats but an older group within the Carnivora, and they independently evolved stabbing canine teeth just like those of *Smilodon*. Nimravids are sometimes called false sabretooths. *False*? Tell that to the early horse *Mesohippus* and the other terrified victims of those giant daggers. Those 'false' sabretooths were living, breathing, snarling, pouncing, probably strong-smelling carnivores, to whose victims they would have seemed anything but false. Another extinct group of 'false sabretooths', the Barbourofelids, lived in the Miocene epoch, later than the Nimravids but earlier than *Smilodon*, and convergently occupying the same niche.

Given that the Carnivora have endowed us with three independently evolved sabretooths at different times in geological history, we might even feel a little let down if there were no marsupial sabretooth. And sure enough, South America rose to the occasion.

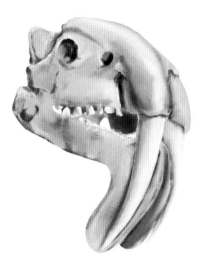
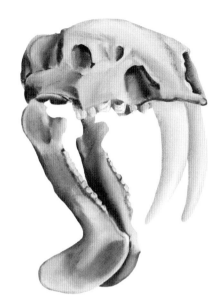

Marsupial sabretooth – Thylacosmilus

The marsupial *Thylacosmilus* looks to have been nearly as formidable as *Smilodon* and the other convergent sabretooths of the Carnivora. On the other hand, it was a bit smaller.

Convergences between animals and human technology can be especially impressive, as we saw in the case of the camera and the vertebrate or octopus eye. Though the discovery was originally thought an outrageous hoax, it is now well accepted that bats hunting by night have their own version – 'echolocation' – of what submariners have converged upon under the name 'sonar' – using echoes of their own sounds to detect targets. Bats are divided into two main groups, the small Microchiroptera, and the large Megachiroptera ('fruit bats' and 'flying foxes'). Microchiropteran bats 'see' with their ears. They have highly sophisticated echolocation, good enough to hunt fast-flying insects. The brain pieces together a detailed model of the world, including insect prey, by a highly sophisticated real-time analysis of the echoes of the bats' own shrieks. When a bat is cruising, its cries just tick over. But when homing in on a moth, which is likely to be taking evasive action, the sounds come out as a rapid-fire stutter like a machine gun. Since each pulse gives the bat an updated picture of the world, machine-gun repetition enables it to cope with a moth's high-speed twists and turns. The higher the pitch, the shorter the wavelength by definition. And only short wavelengths can resolve a detailed picture. That means ultrasound: too high, mostly way too high, for us to hear. Young people can hear the lower end of the bat's frequency range. I nostalgically remember them from my youth as sounding like something between a click and a squeak. We can use instruments called bat detectors, which translate ultrasound into audible clicks.

Slightly less well known is the fact that dolphins and other toothed whales (sperm whales, killer whales) do the same thing, also using ultrasound, and they are up there with bats in sophistication. A more rudimentary form of echolocation has also evolved in shrews, and in cave-nesting birds at least twice independently: in South American oilbirds and Asian cave swiftlets (of bird's-nest soup fame). The birds don't use ultrasound: their cries are low enough for us to hear. Some megachiropterans also use a less precise form of echolocation, but they generate their clicks with their wings rather than with the voice. This too must be seen as yet another convergent evolution of echolocation. One genus of Megachiroptera echolocates using the voice, like Microchiroptera but not so skilfully. Interestingly, molecular evidence indicates that one group of Microchiroptera, the Rhinolophids, are more closely related to Megachiroptera than they are to other Microchiroptera. This would seem to suggest that the Rhinolophids evolved their advanced sonar convergently with the other Microchiroptera. Either that or the majority of Megachiroptera lost it.

Small bats and toothed whales are in a class of their own. Their sonar is of such high quality that 'seeing with their ears' scarcely exaggerates what they do. Echolocation using ultrasound provides them with a detailed picture of their world, which bears comparison with vision. We know this through experimental testing of bats' ability to fly fast between thin wires without hitting them. I have even published the speculation (probably untestable, alas) that bats 'hear in colour'. I stubbornly maintain that it's plausible, because the hues that we perceive are internally generated labels in the brain, whose attachment to particular wavelengths of light is arbitrary. When bat ancestors gave up on eyes, substituting echoes for light, the internal labels for hues would have gone begging, left hanging in the brain with nothing to do. What more natural than to commandeer them as labels for echoes of different quality? I suppose you might call it an early exploitation of what some humans know as 'synaesthesia'.

In one of modern philosophy's most cited papers, Thomas Nagel didactically asked, 'What is it like to be a bat?' One of his points was that we cannot know. My suggestion is that it is perhaps not so very different from what it's like to be us, or another visual animal like a swallow. Pursuing a point from Chapter 1, both swallows and bats build up an internal virtual reality model of their world. The fact that swallows use light, while bats use echoes, to update the model from moment to moment is less important than the nature and purpose of the internal model itself. This is likely to be similar in the two cases, because it is used for a similar purpose – navigation in real time between obstacles, and detection of fast-moving prey. Swallows and bats need a very similar internal model, a three-dimensional one, inhabited by moving insect targets. Both are champion insect hunters on the wing, swallows by day and then, at nightfall, the bats take over. If my speculation is right, the similarity may extend to the use of colours to label objects in the model, even in the case of bats 'seeing with their ears'. Incidentally, each swallow eye has two foveas (regions of special acuity – our eyes have only one, which we use for reading etc.), probably one for distance and one for close vision. Instead of bifocal glasses they have bifocal retinas.

The James Webb Telescope presents us with stunning images of distant nebulae, glowing clouds of red, blue and green. Colour is used to represent wavelength of radiation. But the colours in the photographs are false. They use colour to represent different wavelengths, but they actually lie in the invisible infrared part of the spectrum. And my point is that the brain's convention for representing visible light of different wavelengths is just as arbitrary. One is tempted to feel dissatisfied by false colour images such as those from the James

Webb Telescope: 'But is that *really* what it looks like? Is the telescope telling the *truth*, or are we being fobbed off with false colours?' The answer is that we are always being 'fobbed off' when we look at anything. If you must talk about false colours, everything you ever see – a rose, a sunset, your lover's face – is rendered in the brain's own 'false' colours. Those vivid or pastel hues are internal concoctions manufactured by the brain as coded labels for light of different wavelength. The truth lies in the actual wavelength of electromagnetic radiation. The perceived hue is a fiction, whether it is the false colour rendering of a James Webb photograph, or whether it is the labels that the brain generates to tag the wavelengths of light hitting the retina. My conjecture about bats 'hearing in colour' makes use of the same idea of internally perceived hues being arbitrary labels.

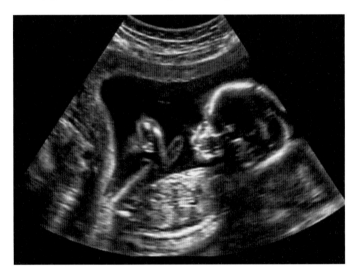

Doctors use ultrasound to 'look' through the body wall of a pregnant woman and see a black-and-white moving image of her developing foetus. The computer uses the ultrasound echoes to piece together an image compatible with our eyes. There is anecdotal evidence that dolphins pay special attention to pregnant women swimming with them. It seems plausible that they are doing with their ears what doctors do with their instruments. If this is so, they could presumably also 'see' inside female dolphins and detect which ones are pregnant. Might this skill be useful to male dolphins choosing mates? No point inseminating a female who is already pregnant.

Bats and dolphins evolved their echo-analysing skills independently of each other. In the family tree of mammals, both are enveloped by relatives who don't do echolocation. A strong convergence, and another powerful demonstration of the power of natural selection. And now for a point that's especially telling for the genetic book of the dead. There's a type of protein called prestin, which is intimately involved in mammal hearing. It's expressed in the cochlea, the snail-shaped hearing organ in the inner ear. As with all biological proteins, the exact sequence of amino acids in prestins is specified by DNA. And, also as is usual, the DNA sequence is not identical in different species. Now here's the interesting point. If you construct a family tree of resemblance based on the genome as a whole, whales and bats are far apart, as you'd expect: their ancestors have been evolving independently of one another since way back in the age of dinosaurs. If, however, you ignore all genes except the prestin gene – if you construct a tree of resemblance based on prestin sequences alone – something remarkable emerges. Dolphins and small bats cluster together with each other. But small bats don't cluster together with non-echolocating large bats, to whom they are much more closely related. And dolphins don't cluster together with baleen whales, which, although related to them, don't echolocate. This suggests that SOF could read the prestin gene of an unknown animal and infer whether it (more precisely its ancestors) lived and hunted in conditions where ultrasonic sonar would be useful: night, dark caves, or other places where eyes are useless, such as the murky water of the Irrawaddy river or the Amazon. I'd like to know whether the two echolocating bird species have bat-like prestins.

This finding on bats and dolphins – the specific resemblance of their prestin genes – strikes me as a pattern for a whole field of future research on the genetic book of the dead. Another example concerns flight surfaces in mammals. Bats fly properly, and marsupial flying phalangers glide, using stretched flaps of skin that catch the air. There's a specific complex of genes, shared by both bats and marsupial phalangers, which is involved in making the skin flaps. It will be interesting to know whether the same genes are shared by the other gliding mammals that we met earlier in this chapter, so-called flying lemurs and the two groups of rodents that independently evolved the gliding habit.

It would be nice to look in the same kind of way at those animals who have returned from land to water – of which whales are only the most extreme example, along with dugongs and manatees. Do returnees to water have genes in common that are not shared by non-aquatic mammals? What other features do they share? Many aquatic mammals and birds have webbed feet. If our hypothetical SOF is presented with an unknown animal who has webbed feet, she can safely 'read' the feet as saying, 'Water in the recent ancestral environment.' But that's obvious. Can we be systematic in our search for less obvious signals of water in the genetic book of the dead? How many other features are diagnostic of aquatic life? Are there some shared genes, such as we saw in the case of prestin for sonar, and skin flaps in bats and sugar gliders? There are probably lots of shared features buried deep in an aquatic animal's physiology and genome. We have just to find them. We can get a sort of negative clue by looking at genes that were made inactive when terrestrial animals took to the water. Just as humans have a large number of smell genes inactivated (see page 53), whale genomes contain several inactivated genes, whose inactivation has been interpreted as beneficial when diving to great depths.

We could proceed along the following lines. We borrow from medical science the technique known as GWAS (genome-wide association study). The idea of GWAS is lucidly and conversationally explained by Francis Collins, former Director of the Human Genome Project, as follows:

What you do for a genome-wide association study is find a lot of people who have the disease, a lot of people who don't, and who are otherwise well matched. And then, searching across the entire genome ... you try to find a place where there is a consistent difference. And if you're successful – and [you've] got to be really careful about the statistics here, so that you don't jump on a lot of false positives – it allows you to zero in on a place in the genome that must be involved in disease risk without having to guess ahead of time what kind of gene you're going to find.

Substitute 'lives in water' for 'disease', and 'species' for 'people', and you have the procedure I am here advocating. Let's call it 'Interspecific GWAS' or IGWAS.

Gather a large number of mammals known to be aquatic. Match each one with a related mammal (the more closely related the better) who lives on land, preferably in dry conditions. We might start with the following list of matched pairs, and the list could be extended.

Water vole	Vole
Water shrew	Shrew
Desman	Mole
Platypus	Echidna
Water tenrec	Land tenrec
Otter	Badger
Seal	Wolf
Yapok	Opossum
Polar bear	Brown bear

To do the IGWAS, you would now look at the genomes of all the animals and try to pinpoint genes shared by the left-hand column and not by the right-hand column. Until all those animals have had their genomes sequenced, and until mathematical techniques are up to the task, proceed with a non-genomic version of IGWAS as follows. Go to work taking measurements of all the animals. Measure all the bones. Weigh the heart, the brain, the kidneys, the lungs, etc., all these weights being expressed relative to total body weight (to correct for absolute size, which is unlikely to be of much interest). By the same token, the bone measurements should be expressed as a proportion of something, just as, in the chelonian example of Chapter 3, the bone lengths were expressed as a proportion of total arm length. Measure the body temperature, blood pressure, the concentrations of particular chemicals in the blood, measure everything you can think of. Some of the measurements might not be continuously varying quantities like centimetres or grams: they might be 'yes or no', 'present or absent', 'true or false'.

Feed all the measurements into a computer. And now for the interesting part. We want to maximise the discrimination between aquatic mammals and their terrestrial opposite numbers. We want to discover which measurements discriminate them, pull them apart. At the same time, we want to identify those features that unite all aquatic mammals, however distantly related from each other. Webbing between the toes will presumably emerge as a good discriminator, but we want to find the non-obvious discriminators, biochemical discriminators, ultimately gene discriminators. Where genomic comparisons are concerned, the GWAS methods already developed for medical purposes will serve. A possible graphic method is a version of the triangular plot of tortoise and turtle limbs that we saw in Chapter 3. Another graphic method is drawing pedigrees with genetic convergences coloured in.

A refinement of IGWAS might order species along an ecological dimension. You could, perhaps, string mammals out along a *dimension of aquaticness*, from whales and dugongs at one extreme to camels, desert foxes, oryxes, and gundis at the other. Seals, otters, yapoks and water voles would be intermediate. Or we might explore a *dimension of arboreality*. We might conclude that a squirrel is a rat who has moved a measurable distance along the dimension of arboreality. Are moles, golden moles and marsupial moles situated at one extreme on a *dimension of fossoriality*. Could we distribute birds along a dimension from flightless cormorants and emus who never fly, at one extreme, to albatrosses at the other, or, even more extreme, to swifts, who even copulate on the wing? Having identified such 'dimensions', could we look for trends in gene frequency as you move along from one extreme to the other? I can immediately foresee alarming complications. The dimensions would interact with other dimensions, and we'd have to call in experts with mathematical wings to fly through multi-dimensional spaces. My own sadly amateur ventures, limited to three dimensions, and using computer simulation rather than mathematics, are in my book *Climbing Mount Improbable*, especially the chapter called 'The Museum of All Shells'.

A group at Carnegie Mellon University in Pittsburgh performed a model example of what I call (they don't) IGWAS. What they studied was not aquaticness but hairlessness in mammals. Most mammals are hairy, and all had hairy ancestors, but if you survey the mammal family tree you notice that hairlessness pops up sporadically among unrelated mammals. See the diagram, which shows a few of the sixty-two species whose genomes were examined.

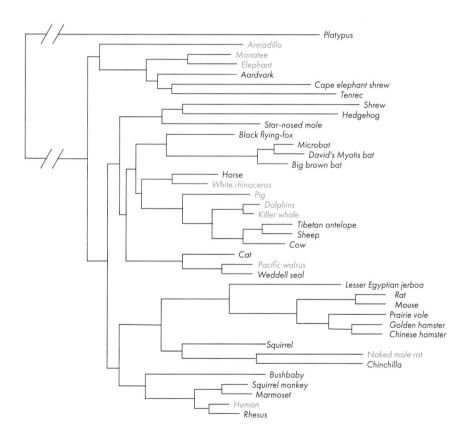

Sporadic distribution of hair loss among mammals

Whales, manatees, pigs, walruses, naked mole rats, and humans have all lost their hair more or less completely (orange names in the diagram). And, which is important, independently of each other in many cases. We can tell this by looking at the hairy closer relatives from among whom they sprang. You remember that echolocating bats and echolocating whales had something else in common – their prestin gene. Do the genomes of the naked species have a gene for hairlessness that they share with each other? The answer is literally no. But only literally. The truth is equally interesting. It turns out that we and other naked species still retain the ancestral genes that make hairs. But the genes have been disabled. And disabled in different ways. What is convergent is the fact of being disabled, but the details are not shared. Incidentally, we again have here a problem for creationists. If an intelligent designer wished to make a naked animal, why would he equip it with genes for making hair and then disable them? Chapter 3 mentions the similar example of the human sense of smell: the olfactory sense genes of our mammal ancestors still lurk within us, but they have been turned off.

One of my favourite examples of convergent evolution is that of weakly electric fish. Two separate groups of fish, Gymnotids in South America and Gymnarchids in Africa, have independently and convergently discovered how to generate electric fields. They have sense organs all along the sides of the body, which can detect distortions that objects in the environment cause in the electric fields. It is a sense of which we can have no awareness. Both groups of fish use it in murky water where vision is impossible. There's just one difficulty. The normal undulating movements typical of fish fatally compromise the analysis of the electric fields measured along the body. It is necessary for the fish's body to maintain a rigid stance. But if their body is rigid, how do they swim? By means of a single longitudinal fin traversing the whole length of the body. The body itself, with its row of electrical sensors, stays rigid, while the single longitudinal fin alone performs the sinuous movements typical of fish locomotion. But there's one revealing difference. In the South American fish, the longitudinal fin runs along the ventral surface, while in the African fish it runs along the back. In both groups of fish, the undulating waves can be thrown into reverse: the fish swim backwards and forwards with apparently equal facility.

The 'duck bill' of the platypus and the huge, flat 'paddle' sticking out of the front end of the paddlefish (Polyodontidae) are both covered with electrical sensors, convergently and independently evolved. In this case the electric fields they pick up are generated, inadvertently, by the muscles of their prey. There is a long-extinct trilobite that also had a huge paddle-like appendage like that of the paddlefish. Its paddle was studded with what look like sense organs, and it seems probable that this represents yet another convergence.

A ringed plover's eggs and chicks lie out on the ground, defenceless except for their camouflage. A fox approaches. The parent is much too small to put up any kind of resistance. So it does an astonishing thing. It attempts to lure the predator away from the nest by offering itself as a bigger prize than the nest. It limps away from the nest, pretending to have a broken wing, simulating easy prey. It flutters pathetically on the ground, wings outstretched, sometimes with one wing stuck incongruously in the air. There's no assumption that it knows what it is doing or why it is doing it (although it may). The minimal assumption we need make is that natural selection has favoured ancestors whose brains were genetically wired up to perform the distraction display, and perfect it over generations. Now, why tell the story in this chapter on convergent evolution? It's because the broken wing display has arisen not once but many times independently in different families of birds. The diagram on the following page is a pedigree of birds, wrapped around in a circle so it fits on the page. Birds who perform the broken wing display are coloured in red, those who don't in blue. You can see that the habit is distributed sporadically around the pedigree, a lovely example of convergent evolution.

My final example of convergence will lead us into the next chapter. More than 200 species, belonging to thirty-six different fish families, practise the 'cleaner' trade. They remove surface parasites and damaged scales from the bodies of larger 'client' fish. Each individual

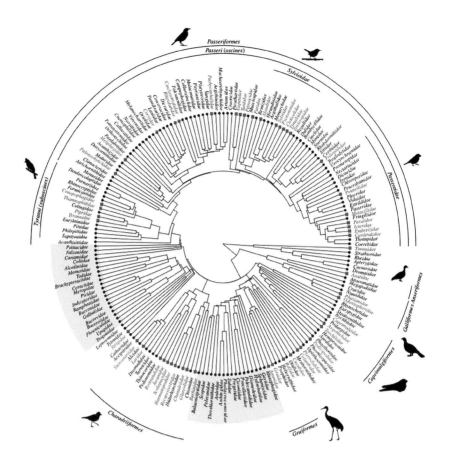

Broken wing display

The remarkable 'broken wing display' crops up again and again in different bird groups (shown in red). Striking testimony to the power of natural selection.

cleaner fish has its own cleaning station, and its own loyal clients who return repeatedly to the same 'barber's shop' on the reef. This site tenacity is important in keeping the benefit exclusively mutual: the cleaner eats the parasites and worn-out scales from the skin of particular client fish, and the client refrains from eating its particular benefactor. Without individual site fidelity, and therefore repeat visits, clients would have no incentive to refrain from eating the cleaner – after being cleaned, of course. Sparing a cleaner would benefit fish in general, including competitors of the sparer. Natural selection doesn't 'care' about general benefit. Quite the contrary. Natural selection cares only about benefit to the individual and its close relations, at the expense of competitors. A bond of individual loyalty between particular cleaner and particular client therefore really matters, and it is achieved by site tenacity. Some cleaners even venture inside the mouth of a client to pick its teeth – and survive to repeat the service on the client's next visit. Cleaner fish advertise their trade and secure their safety by a characteristic dance, often enhanced by a striped pattern – the fishy equivalent of the striped pole insignia of a human barber's shop. This constitutes a safe-conduct pass.

The interesting point for this chapter is that the cleaner habit has evolved many times convergently, not only many times independently in fish but many times in shrimps too. As before, the client fish abide by the covenant and refrain from eating their cleaner shrimps, in just the same respectful way as for cleaner fish. In many cases, cleaner shrimps sport a similar stripe, the 'barber's pole' insignia. It is to the benefit of all that all the 'barber's pole' badges should look similar.

When swimming in the sea, you would be well advised to steer clear of the sharp-toothed jaws of the moray eel. Yet here is a shrimp, calmly picking its teeth. Note, yet again, the red stripe or 'barber's shop pole', telling the moray, 'Don't eat me, I'm your special cleaner. You and I have a mutual relationship. You'll need me again.' Does

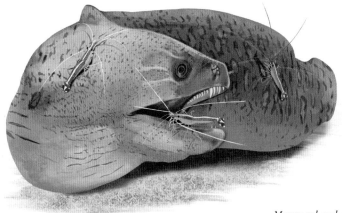

Moray eel and cleaner shrimp

the shrimp feel fear as it trustingly enters those formidable jaws? Does some equivalent of 'trust' pulsate through its cephalic ganglion? I doubt it, but not everyone would agree. Do you?

Not only has the habit evolved independently – convergently – in fish and shrimps. It has evolved convergently many times *within* shrimps, just as it has many times within fish. Even within one family of shrimps, the Palaemonidae, the cleaner trade is practised by sixteen different species, having evolved within the Palaemonidae five times independently. Here's how we know the five evolutions were independent of each other. The method again serves as a model for how we ever know instances of evolution are independent of each other. Look at the family tree of the Palaemonidae, constructed with the aid of molecular genetic sequencing. It contains sixty-eight species of shrimp. Those species that practise the fish-cleaning trade have a little fish symbol by them. There are sixteen species of palaemonid cleaner shrimps. But many of the sixteen cannot be said to have evolved the habit independently. For example, the three species of *Urocardella* are all cleaners, but the picture warns us against counting them as independent: they probably inherited it from their common ancestor.

Six members of the genus *Ancyclomenes* are cleaners, but again we must make the conservative assumption that they inherited it from their common ancestor – and that the habit has been lost in *A.aqabai*, *A.kuboi*, *A.luteomaculatus*, and *A.venustus*. Using this conservative approach, we conclude that the cleaning habit evolved independently in five palaemonid genera but not in all species of those five genera. And the story doesn't end with the Palaemonidae. Two other families of shrimps not shown in the diagram, the Hippolytidae (see moray eel picture above) and the Stenopodidae, also have many species of cleaner.

The Cambridge palaeontologist Simon Conway Morris has treated convergent evolution more vividly and thoroughly than anyone else. In his wittily written *Life's Solution* he points out that convergent evolution is commonly sold as amazing, astounding, uncanny, etc., but there is no need for this. Far from being especially amazing, it's

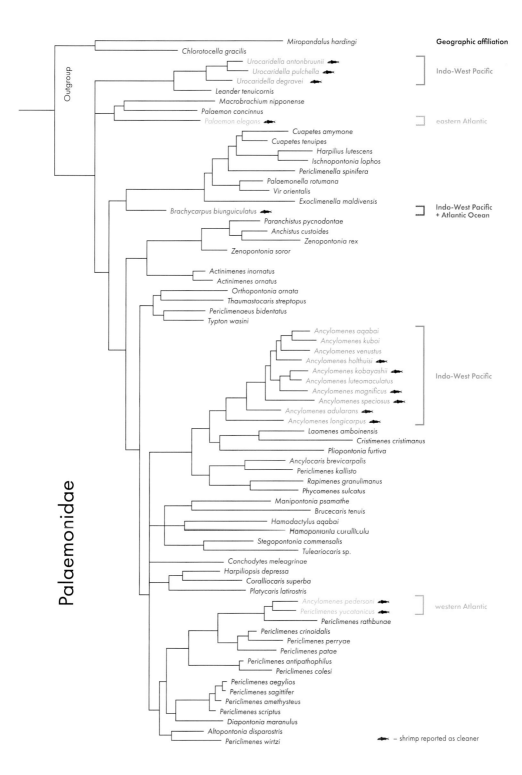

Palaemonidae

Independent evolution of cleaners

exactly what we should expect of natural selection. Convergent evolution is, nevertheless, great for confounding armchair philosophers and others who underestimate the power of natural selection and the magnificence of its productions. In addition to 110 densely packed pages of massively researched endnotes and references to the biological literature, *Life's Solution* has three indexes: a general index, a name index and – this must surely be unique – a 'convergences index'. It runs to five double-column pages and around 2,000 examples of convergence. Of course, not all of them are as impressive as the pill-bugs, the moles, the gliders, the sabretooths, or the fish-cleaners but even so …

Convergent evolution can be so impressive, it makes you wonder how we know the resemblance really is convergent. That's the power of natural selection, the immense yet subtle power that underpins the whole idea of the genetic book of the dead. Pill woodlouse and pill millipede, alike as two pills, how do we know one is a crustacean, the other a distant myriapod? There are numerous tell-tale clues. The deep layers of the palimpsest are never completely over-written. The glyphs of history keep breaking through. And, if all else fails, molecular genetics cannot be denied.

Convergence of animals with widely separated histories is one manifestation of the power of selection to write layer upon layer of the palimpsest. Another is its converse: evolutionary *divergence* from a common historic origin, natural selection seizing a basic design and moulding and twisting it into an often bizarre range of functionally important shapes. The next chapter goes there.

6

Variations on a Theme

As we saw in Chapter 3, molecular comparison conclusively shows that whales are located deep within the even-toed ungulates, the artiodactyls. By 'located deep within', I mean something very specific and surprising. It's worth repeating. We're talking about much more than just a shared ancestor, with the whales going one way, and the artiodactyls the other. That would not have been surprising. 'Deep within' means that some artiodactyls (hippos) share a more recent ancestor with whales than they share with the rest of the artiodactyls whom they much more strongly resemble. This has been known for more than twenty years, but I still find it almost incredible, so overwhelming is the submersion under surface layers of palimpsest. Of course, this doesn't mean whales' ancestors were hippos or even resembled hippos. But whales are hippos' closest living relatives.

What is it that's so special about whales, so special that new writings in their book of the dead so comprehensively obliterated almost every trace of that earlier world, of grazing prairies and galloping feet, which must lie buried far down in the palimpsest? How did the whales manage to diverge so completely from the rest of the artiodactyls? How were they able so comprehensively to escape their artiodactyl heritage?

The answer probably lies in that word 'escape'. Cattle, pigs, antelopes, sheep, deer, giraffes, and camels are relentlessly disciplined by gravity. Even hippos spend significant amounts of time on land, and indeed can accelerate their ungainly bulk to an alarming speed. The land-dwelling artiodactyl ancestors of whales had to submit to gravity. In order to move, land mammals must have legs stout enough to bear their weight. A land animal as big as a blue whale would need legs half way to Stonehenge pillars, and it'd have a hard time surviving, with heart and lungs smothered suffocatingly by the body's own weight. But in the sea, whales shook off gravity's tyranny. The density of a mammal body is approximately that of water. Gravity never goes away, but buoyancy tames it. When their artiodactyl ancestors took to the water, whales shed the need for leggy support, and the fossil evidence beautifully lays out the intermediate stages.

A major milestone marks the point where, like dugongs and manatees but unlike seals and turtles, whales gave up returning to land even to reproduce. That was the final release from gravity, as buoyancy totally took over. Whales were free to grow to prodigious size, literally insupportable size. A whale is what happens when you take an ungulate, cut it adrift from the land and liberate it from gravity. All manner of other modifications followed in the wake of the great emancipation, and they richly defaced the ancient palimpsest. Forelegs became flippers, hind limbs disappeared inside and shrank to tiny relics, the nostrils moved to the top of the head, two massive horizontal flukes – lobes stiffened not by bone but by dense fibrous tissue – sprouted sideways to form the propulsive organ. Numerous profound alterations of physiology and biochemistry allowed deep diving, and hugely prolonged intervals between breaths. Whales switched from a (presumed) herbivorous diet to one dominated by fish, squid, and – in the case of the baleen whales – filtered shoals of krill in lavish quantities.

Fish, too, are allowed by buoyancy to adopt bizarre shapes (see pictures on pages 122–123), which gravity on land would forbid. In the case of teleost (bony as opposed to cartilaginous) fish, the buoyancy is perfect, owing to that exquisite device, the swim-bladder, buried deep within the body. By manipulating the amount of gas in the swim-bladder, the fish is able to adjust its specific gravity and achieve perfect equilibrium at whatever happens to be its preferred depth at any time.

I think that's what makes a home aquarium such a restful furnishing for a room. You can dream of drifting effortlessly through life, as a fish drifts through water in perpetual equilibrium. And it is the same hydrostatic equilibrium that frees fish to assume such an extravaganza of shapes. The leafy sea dragon trails clouds of glorious fronds, and you feel you could almost identify the species of wrack that those fronds mimic. You must peer deep between them to discern that they are parts of a fish: a modified sea horse – which is itself a distorted caricature of the 'standard fish' design of more familiar cousins such as trout and mackerel.

Most predatory fish actively seek and pursue prey, and this expends a considerable proportion of the energy obtained from the food caught. Angler fish, of which there are several hundred species sitting on the sea bottom, save energy by luring prey to come to them. The anglers themselves are superbly camouflaged. A fishing rod (modified fin spine) sprouts from the head. At its tip is a lure or bait, which the angler fish waves around in a tempting manner. Unsuspecting prey are attracted to the bait, whereupon the angler opens its enormous mouth and engulfs the prey. Different species of angler favour different baits. With some it resembles a worm, and it jiggles about plausibly as the angler waves its rod. Angler fish of the dark deep sea harbour luminescent bacteria in the tip of the rod. The resultant glowing lure is very attractive to other deep-sea fish, and invertebrate prey such as shrimps. Convergently, snapping turtles rest with their mouth open, wiggling their tongue like a worm, as bait for unsuspecting prey fish.

Sea horses and angler fish are extreme exponents of the adaptive radiation of teleost fish. They also, in their different ways, sport unusual sex lives. The sex life of angler fish is nothing short of bizarre. Everything I said in the previous paragraph applies to female angler fish only. The males are tiny 'dwarf males', hundreds of times smaller

Lionfish

Weedy sea dragon

Leafy sea dragon

Marlin

Trumpet fish

Sunfish

than females. A female releases a chemical, which attracts a dwarf male. He sinks his jaws into her body, then digests his own front end, which becomes buried in the female's body. He becomes no more than a small protuberance on her, housing male gonads from which she extracts sperm when she needs to. It is as though she becomes a hermaphrodite, except that 'her' testes possess a different genotype from her own, having invaded from outside in the form of the dwarf male locked into her skin.

Many species of fish are livebearers – females get pregnant like mammals and give birth to live young. Sea horses are unusual in that

Gulper eel

Seahorse

Puffer

Sloane's viper fish

Ghost pipefish

Angler fish

Freed by buoyancy from the constraints of gravity, fish were able to evolve an astonishing variety of shapes

it's the male who gets pregnant, carries the young in a belly pouch, and eventually gives birth to them. Do you wonder, then, how we define him as male? Throughout the animal and plant kingdoms, the male sex is easily defined as the one that produces lots of small gametes, sperms, as opposed to fewer, larger, eggs.

Adaptive radiation means evolutionary divergence fanning out from a single origin. It is seen in an especially dramatic way when new territory suddenly becomes available. When, 66 million years ago, a celestial catastrophe cleared 76 per cent of all species from the planet, the stage was wide open for mammalian understudies to step into the dinosaurs' vacated costumes. The subsequent adaptive radiation of mammals was spectacular. From the small, burrowing creatures who survived the devastation, probably by hibernating in safe little underground bunkers, a comprehensive range of descendants, ranging hugely in size and habit, appeared in surprisingly quick time.

On a smaller scale and a much shorter timescale, a volcanic island can spring up suddenly (suddenly by the standards of geological time) through volcanic upwelling from the bottom of the sea. For animals and plants it is virgin territory, barren, untenanted, open to exploitation afresh. Slowly (by the standards of a human lifetime) the volcanic rock crumbles and starts to make soil. Seeds fly in on the wind, or are transported by birds and fertilised with their droppings. From being a black lava desert, the island greens. Winged insects waft in, and tiny spiders parachuting under floating threads of silk. Migrating birds are blown off course, land for recuperation, stay, reproduce; their descendants evolve. Fragments of mangrove drift in from the mainland, and the occasional tree uprooted by a hurricane. Such freak raftings carry stowaways – iguanas, for instance. Step by accidental step, the island is colonised. And then descendants of the colonists evolve, rapidly by geological standards, diversifying to fill the various empty niches. Diversification is especially rich in archipelagos, where driftings between islands happen more frequently than from the mainland to the archipelago. Galapagos and Hawaii are textbook examples.

A volcano is not the only way new virgin territory for evolution can open up. A new lake can do it too. Lake Victoria, largest lake in the tropics and larger than all but one of the American Great Lakes, is extremely young. Estimates range from 100,000 years to a carbon-dated figure of only 12,400 years. The discrepancy is easily explained. Geological evidence shows that the lake basin formed about 100,000 years ago, but the lake itself has dried up completely and refilled several times. The figure of 12,400 years represents the age of the latest refilling, and therefore the age of the current lake in its large geography. And now, here is the astonishing fact.

There are about 400 species of Cichlid (pronounced 'sicklid') fish

Nimbochromis livingstonii   *Lamprologus lemairii*

in Lake Victoria, and they are all descended from probably as few as two founder lineages that arrived from rivers within the short time that the lake has existed. The same thing happened earlier in the other great lakes of Africa, the much deeper Lakes Tanganyika and Malawi. Each of the three lakes has its own unique radiation of Cichlid fishes, different from, but parallel to, the others.

Here's a slightly macabre example of this parallelism. In Lake Malawi (where I spent my earliest bucket-and-spade beach holidays), there is a predatory fish called *Nimbochromis livingstonii*. It lies on the bottom of the lake pretending to be dead. It even has light and dark blotches all over its body, giving the appearance of decomposition. Deceived into boldness, small fish approach to nibble at the corpse, whereupon the 'corpse' suddenly springs into action and devours the small fish. This hunting technique was thought to be unique in the animal kingdom. But then exactly the same trick was discovered in Lake Tanganyika, the other great Rift Valley lake. Another Cichlid fish, *Lamprologus lemairii*, has independently, convergently, hit upon the same death-shamming trick. And it has the same blotchy appearance, suggestive of death and decay. In both lakes, adaptive radiation independently hit upon the same somewhat gruesome way of getting food. Along with dozens of other ways of life, independently discovered in parallel in the two similar lakes.

My old friend, the late George Barlow, vividly described the three great lakes of Africa as Cichlid factories. His book, *The Cichlid Fishes*, makes fascinating reading. The Cichlids have so much to teach us about evolution in general and adaptive radiation in particular. Each of the three great lakes has its own, independently evolved radiation of several hundred Cichlid species. All three lakes tell the same story of explosive Cichlid evolution, yet the three histories unfolded entirely independently. All three began with a founder population of very few species. Each of the three followed a parallel evolutionary course of massive radiation into a huge variety of 'trades' or ways of life – the same great range of trades being independently discovered in all three lakes.

You might think the oldest lake would have the most species. After all, it's had the longest time to evolve them. But no. Lake Tanganyika, easily the oldest at about 6 million years, has only (only!) 300 species. Victoria, a baby of only 100,000 years, has about 400 species. Lake Malawi, intermediate in age at between 1 and 2 million years, has the largest species count, probably around 500, although some estimates exceed 1,000. Moreover, the size of the radiation seems unrelated to the number of founder species. The huge radiations in Victoria and Malawi trace back substantially to only one lineage of Cichlids, the Haplochromines. The relatively venerable Lake Tanganyika's approximately 300 species appear to stem from twelve different founder lineages, of which the Haplochromines are only one.

What all this suggests is that young Lake Victoria's dramatic explosion of species is the model for all three lakes. All three probably took only tens of thousands of years to generate several hundred species. After the explosive beginning, the typical pattern is probably to stabilise the number, or it may even decrease, such that the final number of species is not correlated with the age of the lake, or with the number of founder species. The Cichlids of Lake Victoria show how fast evolution can proceed when it dons its running shoes. We cannot expect that such an explosive rate is typical of animals in general. Think of it as an upper bound.

And when you work it out, even Lake Victoria's feat is not quite so surprising as first appears. Although the lake in its present form is only some 12,400 years old, I've already mentioned that a lake filled the same shallow basin 100,000 years ago. In the intervening years it has largely dried up several times and refilled, the latest such episode occuring with the refill of 12,400 years ago. Lake Malawi shows how dramatically these lake levels can fall and rise. Between the fourteenth and nineteenth centuries, the water level was more than 100 metres lower than today. Unlike Lake Victoria, however, it came nowhere close to drying up altogether. In its Rift Valley chasm, it is nearly ten times as deep as Victoria. In shallow Lake Victoria, as each drying cycle occurred, the lowering of the water level would have left numerous ponds and small lakes, these becoming reunited at the next iteration of the refill cycle. The temporary isolation of the fish trapped in the residual ponds and small lakes enabled them to evolve separately – no gene flow between ponds. At the next refill of the cycle, they were reunited, but by then they would have drifted apart genetically, too far to interbreed with those who had been stranded in other ponds. If this is correct, the drying/refilling alternation provided ideal conditions for speciation (the technical term for the evolutionary origin of a new species, by splitting of an existing species). And it means that, from an evolutionary point of view, we could regard the true age of Lake Victoria as 100,000 years, not 12,400. Still very young.

Given 100,000 years to play with, what sort of interval between speciation events would yield 400 species, starting, hypothetically, with a single founding species? Is 100,000 years long enough? Here's how a mathematician might reason: a back-of-the-envelope calculation, making conservative assumptions throughout, to be on the safe side. There are two extremes, two bounds bracketing the possible rate of speciation, depending on the pattern of splitting. The most prolific pattern (an improbable extreme) is where every species splits into two, yielding two daughter species which, in turn, split into two. This pattern yields exponential growth of species numbers. It would take only between eight and nine speciation cycles to yield 400 species (2^9 is 512). An interval of 11,000 years between speciations would do the trick. The least prolific pattern (also an improbable extreme) is where the founder species 'stays put' and successively throws off one daughter species after another. This would require far more speciation events, about 400, to reach the tally of 400 species: a speciation event every 250 years. How to estimate a realistic intermediate between these two extremes? A simple average (arithmetic mean) gives an estimate of between 5,000 and 6,000 years between speciations, which is enough time. Our mathematician, however, might be more cautious and recommend the geometric mean (multiply the two numbers together and take the square root). One reason to prefer it is that it captures the stronger influence of an occasional very bad year. This more conservative estimate asks for an interval of about 1,600 years between speciations. Somewhere between the two estimates is plausible, but let's bend over backwards to be cautious and use the estimate of 1,600 years. Cichlid fish typically reach sexual maturity in under two years, so let's again be conservative and assume a two-year generation time. Then we'd need about 800 fish generations between speciation events, in order to generate 400 species in 100,000 years. Eight hundred generations is enough for plenty of evolutionary change.

How do I know 800 generations is plenty of time? Again, mathematicians can do back-of-the-envelope calculations to assist intuition. One calculation that I like was done by the American botanist Ledyard Stebbins. Imagine that natural selection is driving mouse-sized animals towards larger size. Stebbins, too, bent over backwards to be conservative, by assuming a very weak selection pressure, so weak that it could not be detected by scientists working in the field, trapping mice and measuring them. In other words, natural selection in favour of larger size is assumed to exist but to be so slight and subtle that it is below the threshold of detectability by field researchers. If the same undetectably weak selection pressure were maintained consistently, how long would it take for the mice to evolve to the size of an elephant? The answer Stebbins calculated was about 20,000 generations, the blink of an eye by geological standards. Admittedly, it's a lot more than our 800 generations, but we weren't talking about anything so grandiose as mice turning into elephants. We were only talking about Cichlid fishes changing enough to be incapable of interbreeding with other species. Moreover, Stebbins's assumptions, like ours, were conservative. He assumed a selection pressure so weak that you couldn't measure it. Selection pressures have actually been measured in the wild, for example on butterflies. Not only are they easily detectable, they are orders of magnitude stronger than the sub-threshold, under-the-radar pressure assumed by Stebbins. I conclude that 100,000 years is a comfortably long time in Cichlid evolution, easily enough time for an ancestral species to diversify into 400 separate species. That's fortunate, because it happened!

Incidentally, Stebbins's calculation is an instructive antidote to sceptics who think geological time is not long enough to accommodate the amount of evolutionary change we observe. His 20,000 generations to wreak the change from mouse to elephant is so short that it would ordinarily not be measurable by the dating methods of geologists. In other words, a selection pressure too weak to be detectable by field geneticists is capable of yielding major evolutionary change so fast that it could look instantaneous to geologists.

The crustaceans are another great group of mostly aquatic animals with spectacular evolutionary radiations, from much more ancient common sources. In this case, it is the modification of a shared anatomy that impresses. Rigid skeletons permit movement only if built up of hinged units, bones in the case of vertebrates, armoured tubes and casings in the case of crustaceans and other arthropods. Because these bones and tubes are rigid and articulated, there is a finite number of them, each one a unit that can be named and recognised across species. The fact that all mammals have almost the same repertoire of nameable bones (206 in humans) makes it easy to recognise evolved differences as distortions of each named bone: ulna, femur, clavicle, etc. The same is true of crustacean skeletal elements, with the bonus that, unlike bones, they are externally visible.

The great Scottish zoologist D'Arcy Thompson took six species of crab and looked at just one unit of the skeleton, the main portion of the body armour, the carapace, of each.

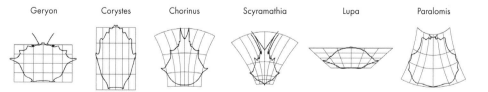

Geryon Corystes Chorinus Scyramathia Lupa Paralomis

He arbitrarily chose one of the six, it happened to be *Geryon* (far left), and drew it on a rectangular grid. He then showed that he could approximate the shape of each of the other five, simply by distorting the grid in a mathematically lawful way. Think of it as drawing one crab on a sheet of stretched rubber, then distorting the rubber sheet in mathematically specified directions to simulate five other shapes. These distortions are not evolutionary changes. The six species are all contemporary. No one species is ancestral to any other, they share ancestors who are no longer with us. But they show how easily changes in embryonic development (altered gradients of growth rates, for instance) can yield an illuminating variety of crustacean form with respect to one part of the exoskeleton. D'Arcy Thompson did the same thing with many other skeletal elements including human and other ape skulls.

Of course, bodies are not drawn on anything equivalent to stretched rubber. Each individual develops afresh from a fertilised egg. But changes in growth rates, of each part of the developing embryo, can end up looking like the distortions of stretched rubber. Julian Huxley applied D'Arcy Thompson's method to the relative growth of different body parts in the developing embryo. Such embryological changes are under genetic control, and evolutionary changes in gene frequencies generate evolutionary variety, again looking like stretched rubber. And of course it isn't just the carapace. The same kind of evolutionary distortion is seen in all the elements of the crustacean body (and the bodies of all animals but often less obviously). You can see how the same parts are present in each specimen, just emphasised to different degrees. The differential emphasis is achieved by different growth rates in different parts of the embryo.

Crustaceans are exceedingly numerous. With characteristic wit, the Australian ecologist Robert May said, 'To a first approximation, all species are insects,' yet it has been calculated that there are more individual copepods (crustacean water fleas) than there are individual insects in the world. The painting opposite, by the zoologist Ernst Haeckel (1834–1919), Darwin's leading champion in Germany, is a dazzling display of the anatomical versatility of the copepods.

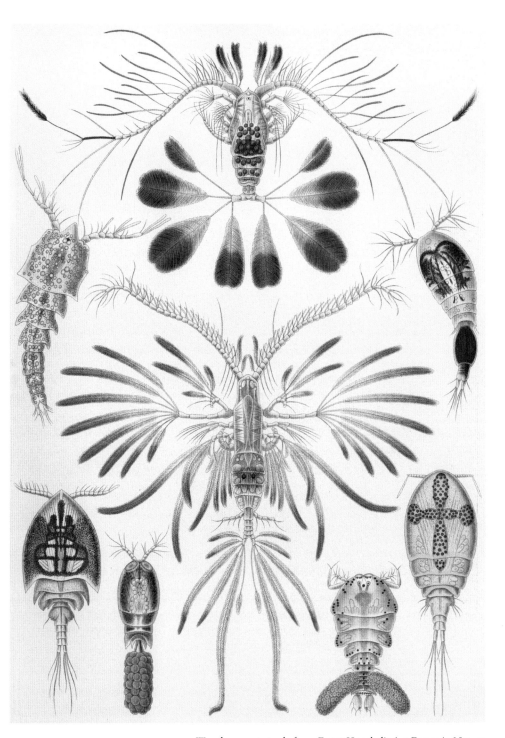

Wondrous copepods from Ernst Haeckel's Art Forms in Nature

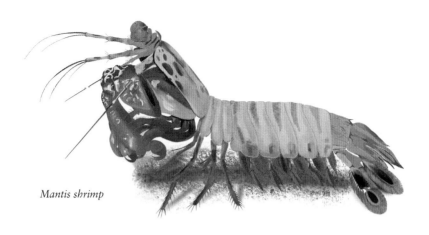

Mantis shrimp

Here's a typical adult crustacean, a mantis shrimp. Well, mantis shrimps (Stomatopods) are typical with respect to their body plan, which, together with their colourful beauty, is why I've chosen one for this purpose. But they include some formidable customers who are far from typical in one alarming respect. They pack a punch, literally. With vicious blows from club-like claws, they smash mollusc shells in nature, while in captivity the blow from a large smasher, travelling as fast as a small-calibre rifle bullet, will shatter the glass of your aquarium tank. The energy released is so great that the water boils locally and there is a flash of light. You don't want to mess with a mantis shrimp, but they're a wonderful example of the diverse modification of the basic crustacean body plan.

Mantis shrimps are not to be confused with the (literally) stunning 'pistol shrimps' or 'snapping shrimps' (Alpheidae), who in their way also beautifully illustrate the diversity of crustacea. These have one enlarged claw, somewhat bigger than the other. They snap the enlarged claw with terrific force, generating a shock wave – a violent pulse of extreme high pressure immediately followed by extreme low pressure in its wake. The shock wave stuns or kills prey. The noise is among the loudest heard in the sea, comparable to the bellows and squeaks of large whales. Muscles are too slow to generate high-speed movement such as the snapping claws of pistol shrimps or the punching clubs of mantis shrimps (or indeed the jump of a flea). They store energy in an elastic material or spring, and then suddenly release it – the catapult or bow-and-arrow principle.

Crustacea dazzle with diversity. But it is a constrained diversity. To repeat the point, which is the reason I chose crustaceans for this chapter, you can in every species easily recognise the same parts. They are connected to each other in the same order, while differing hugely in shape and size. The first thing you notice about the basic crustacean body plan is that it is segmented. The segments are arrayed from front to rear like a goods train with trucks (American freight train with wagons or cars). The segmentation of centipedes and millipedes is even more obviously train-like because most of their segments are the same. A mantis shrimp or a lobster is like a train whose trucks are the same in a few respects (wheels, bogies, and coupling hooks, say) but different in other ways (cattle wagons, milk tanks, timber carriers, etc.).

Crustaceans in their evolution achieve astonishing variety by changing the trucks over evolutionary time, while never losing sight of the train. Varied as they are, the segments of a mantis shrimp are still visibly a train built to the same pattern as any other crustacean, each bearing a pair of limbs that fork at the tip. The claw of a crab or lobster is a conspicuous example of the fork. As you move from front to rear of the animal, the paired appendages consist of antennae, various kinds of mouth parts, claws, then four pairs of legs. Move backwards further, and the segments of a lobster or mantis shrimp's abdomen each have small, jointed appendages called swimmerets underneath, on both sides, each often ending in a little paddle. In a lobster or, even more so, a crab, the segments of the thorax and head are hidden beneath a shared cover, the carapace. But their segmentation is betrayed by the appendages, walking legs in the case of four of them, antennae, large claws and mouth parts at the front end. The rear end of the abdomen, the guard's van (American caboose) of the train, has a special pair of flattened appendages called uropods. When I first visited Australia, I was intrigued to see, laid out in a buffet, what they call bay bugs. These have what look like uropods at the front end as well as the rear, a sort of crustacean version of Doctor Dolittle's Pushmi-Pullyu, but with two rear ends instead of two heads. This is not all that surprising, as we shall now see.

The segmentation of arthropods and vertebrates was once thought to have evolved independently. No longer, and thereby hangs a fascinating tale, a tale that is true too of other segmented animals such as annelid worms. Just as the segments are arrayed in series from front to rear like a train, so the genes controlling the segments are arrayed in series along the length of a chromosome. This revolutionary discovery overturned the whole attitude to zoology that I had learned as a student, and I find it wonderful. To pursue the railway analogy, there's a train of gene trucks in the chromosome to parallel the train of segment trucks in the body.

It's been known for more than a century that mutant fruit flies can have a leg growing where an antenna ought to be. That mutation is called *antennapedia* for obvious reasons, and it breeds true. There are other dramatic mutations in fruit flies, for example *bithorax*, which has four wings like normal insects, instead of the two-winged pattern that gives flies their name, Diptera. These major mutations are all explained by changes in the sequentially arranged genes in the 'chromosome train'. When I first saw that bay bug in a Great Barrier Reef restaurant, I immediately wondered whether bay bugs had originally evolved by a mutation similar to *antennapedia*, in this case duplicating uropods at the front end of the animal.

This kind of effect has been neatly shown by Nipam Patel and his colleagues. They work on a marine crustacean called *Parhyale*, belonging to the Amphipod order. I remember being fascinated by the hundreds of small amphipods in the cold stream on our farm, in the course of which my parents dug out a pool for us to swim. The swarms of exuberantly jumping 'sandhoppers' that we so often encounter on beaches are another familiar example. We met isopods, in the flattened shape of 'pill bugs', in the previous chapter. Amphipods are different. They are flattened left to right rather than back to belly. And, in *Parahyale* and many others, their appendages are far from all the same. Some of their legs point in what seems to be the 'wrong' direction. Three of the 'trucks' appear to be 'coupled' up backwards (red shading in left picture on the next page). Patel and his colleagues, by means of ingenious manipulations of

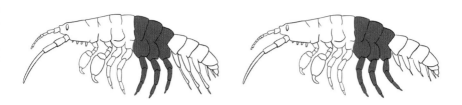

the genes controlling the trucks of the train, were able to change the three reversed segments, coupling the trucks so that all the limbs faced in the same direction (right picture). The way this works is that the three backwards segments are replaced by duplicates of the three segments in front of them. The Patel group achieved equally interesting manipulations of other segments but the work, though fascinatingly ingenious, would take us too far afield.

We vertebrates too are segmented, but in a different way. This is obvious in fish, and it remains pretty clear in our ribs and vertebral column. Snakes carry it to an extreme – sort of like centipedes but with internal ribs instead of external legs. We now understand the embryological mechanism whereby segments are multiplied up. Surprisingly, actually rather wonderfully, it has turned out to be pretty much the same in vertebrates and arthropods. Hence, we understand how it is that different snake species evolve radically different numbers of vertebrae ranging from around 100 to more than 400 – compared to our thirty-three. Vertebrae, whether or not they sprout ribs, all have similar coupling mechanisms to the neighbouring 'trucks of the train', and all have similar blood vessels, and sensory and motor nerves, connected to the spinal cord, which passes through them. As I just mentioned, one of the most revolutionary discoveries of recent zoology is that the embryological mechanisms underlying segmentation in arthropods and vertebrates, deep in the lower levels of their palimpsests, are tantalisingly similar. Once again, the truly beautiful fact is that in both groups, genes are laid out along chromosomes in the same order as the segments that they influence.

Although crustaceans all follow the segmented plan boldly written in the depths of the palimpsest, the 'trucks' vary so extravagantly that the simile of the train can become rather strained. Sometimes many of the segments join together to form a singular body, as in crabs. Often the appendages sprouting from the segments vary spectacularly, ranging from the formidable claws near the front of a lobster, or the punching clubs of a mantis shrimp, to the swimmerets arrayed under the abdomen. Crustaceans range in size from 'water fleas' at less than 1 millimetre to the Japanese spider crab *Macrocheira* with a limb span that can reach 3 metres (10 feet). Frightening as this creature might be to meet, it is harmless to humans. Imagine the handshake of a lobster, or the punch of a mantis shrimp, that size!

Japanese spider crab

Crabs can be thought of as lobsters with a truncated tail (abdomen) curled up under the main body, so you don't see it unless you upend the animal. The crab abdomen bears a passing resemblance to the ape/human coccyx, both being made of a handful of segments from an ancestral tail squashed up. Hermit crabs are strictly not crabs, but belong in their own group (Anomura) within the crustacea. Their abdomen is not squashed up underneath them as in true crabs, but soft and curled round to one side, to fit the discarded mollusc shells that hermit crabs inhabit. The process by which they choose their shells, and compete with one another for favoured shells, is fascinating in its own right. But that's another story. In this chapter they serve as yet another illustration of the wonderful diversity of crustaceans.

The larvae of crustaceans show the group's diversity at least as gloriously as the adults. But still the basic train design is palpable throughout. Perhaps even more dramatically than in the case of adult crustaceans, it is as though natural selection pulled, pushed, kneaded, or distorted the various segments of the body with wild abandon. Different species of crustacean pass through nameable larval stages, free-living animals in their own right, often leading a very different life from the adults – as caterpillars live very differently from butterflies among the insects. The zoea is one such larval type. It is the last stage before the adult, in crabs, lobsters, crayfish, shrimps, bay bugs, and their kind – the decapod crustaceans.

Overleaf is a page full of assorted zoeas to show how easily the basic crustacean plan can be stretched and bent around in evolution, as though made of modelling clay. What I take away from these exquisite little creatures is that all have the same parts, they just vary the relative sizes and shapes of those parts. They all look like distorted versions of each other. That's what evolutionary diversification is all about, and the crustacea show it as plainly as any animal group. You can match up the corresponding parts in all the species, and can clearly see how the different species have pulled, stretched, twisted, swelled, or shrunk the same parts in different ways over evolutionary time. It is wondrous to behold, you surely agree.

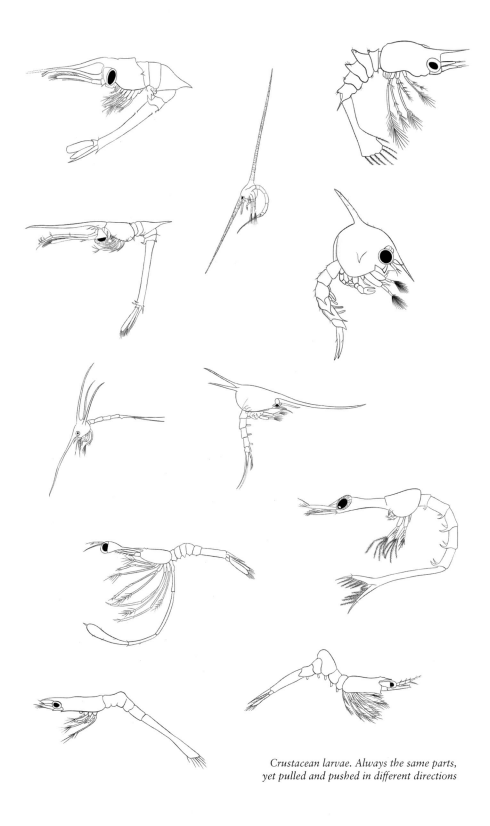

Crustacean larvae. Always the same parts,
yet pulled and pushed in different directions

Zoeas may look a little like the adults they are to become. But they need to survive in a very different world, usually the world of plankton, and their bodies are versatile enough to evolve into all sorts of unlikely distortions – written in surface layers of the palimpsest. Many of them sport long spikes, presumably to make them difficult to swallow. The impressive spikes of the planktonic zoea at top middle are nowhere to be seen in the typical adult crab it is to become. Truth be told, the adult in this case is not easily seen at all under the sea urchin that it habitually carries around on its back – presumably to gain protection via the urchin's own spikes. Notice the long, prominent abdomen of the larva, with its easily discerned segments. As with all crabs, the adult abdomen is neither long nor prominent but tucked discreetly under the thorax.

An earlier larval stage than the zoea, found in most crustacean life cycles, is the nauplius larva. Unlike zoeas, which bear some sort of resemblance to the adult they will become, naupliuses have an appearance all their own. There's another larval stage possessed by some crustaceans, the cyprid larva, presumably so called because it resembles the adult of a water flea called *Cypris*. Perhaps the adult *Cypris* is an example of the overgrown larva phenomenon, which is a fairly common way for evolution to progress. Below is the cyprid larva of a member of the rather obscure crustacean sub-class, Facetotecta.

This larva is unmistakeably crustacean, with a head shield, and abdominal segments bearing typically crustacean forked appendages. From 1899, when the larvae were first discovered, until 2008, nobody knew what adult facetotectans looked like. And they still have never been seen in the wild. What happened in 2008 was that a group of experimentalists succeeded in persuading larvae to turn into a

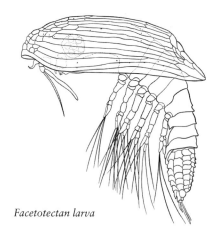

Facetotectan larva

precursor of the adult. They did it by means of hormone treatment. The subtitle of their paper is 'Towards a solution to a 100-year-old riddle'. The adults turn out to be soft, unarmoured, slug-like or worm-like creatures with no visible segments and no appendages, presumably parasites, although nobody knows who their victims are. You wouldn't know, to look at them, that they are crustaceans at all. This experiment recalls a similar one by Julian Huxley with axolotls in 1920. Axolotls are vertebrates, members of the Amphibia. They look like tadpoles; indeed they are tadpoles, but sexually mature tadpoles, and they reproduce. They evolved from larvae who would once have turned into salamanders. The adult stage of their life history was cut off during their evolution, as the larvae became sexually capable. By treating them with thyroid hormone, Julian Huxley succeeded in turning them into the salamanders that their ancestors once were. This experiment may have inspired his younger brother Aldous Huxley to write his novel *After Many a Summer*, in which an eighteenth-century aristocrat discovered how to cheat death – and developed, 200 years later, into a shaggy, long-armed ape humming a Mozart aria. We humans are 'larval' apes!

Those slug-like facetotectans are yet another manifestation of crustacean diversity. They must be descended from adults who had segments and limbs like any respectable crustacean. But the most characteristically crustacean scripts of the palimpsest have been almost completely obliterated by parasitic over-writing, while being retained in the larva. Degenerative evolution of this kind is common in parasites hailing from many parts of the animal kingdom. Within the crustacea, it is also shown to an extreme in certain members of the barnacle family, though not the typical barnacles that encrust rocks at the seaside and prick your bare feet when you walk on them.

As a boy on a seaside holiday, I remember being frankly incredulous when my father told me barnacles are really crustaceans. I thought they were molluscs because, well, they look like molluscs. Nothing like crustaceans, anyway, until you look carefully inside. The barnacles that cling close to rocks look like miniature limpets, while goose barnacles look like mussels on stalks. So how do we know they are

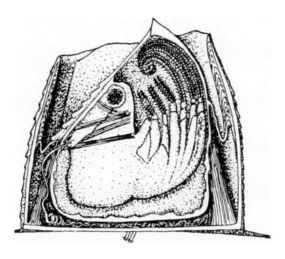

really crustaceans? Look inside. Or see Darwin's own drawing above and you find a shrimp-like creature lying on its back and sweeping the water with its comb-like limbs to filter out swimming morsels of food. As we have by now come to expect, the larvae of barnacles are more unmistakeably crustacean than the adults. Before the adult settles down to its sedentary permanence, it is a free-swimming larva in the plankton. On the left is the nauplius larva of *Semibalanus*, a small rock barnacle with, for comparison, the nauplius larva of a shrimp, *Sicyonia*.

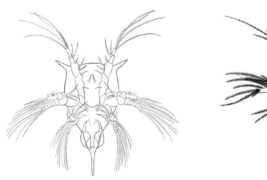

Barnacle larva

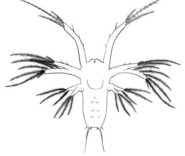

Shrimp larva

Barnacles don't encrust only rocks. To a barnacle, a whale would seem like a gigantic mobile rock. Not surprisingly, some barnacles make their home on the surface of whales, and there are species of barnacle who live nowhere else. Others ride on crabs, and some of them, especially *Sacculina*, evolved into the most extreme examples of divergence from normal crustacean form. They moved, in evolutionary time, from the outside of the crabs to the inside, and became internal parasites bearing no apparent resemblance to a barnacle – or even any kind of animal. Parasites often evolve in a direction that could fairly be called degeneration, and *Sacculina* is an extreme example of this. I shall return to it in the final chapter.

There are many groups of animals that I could have chosen to illustrate evolutionary divergence and variation on a theme. Fish and crustaceans do it perhaps more spectacularly than any other groups, and I chose especially the larvae of crustaceans, partly because, living in the plankton as most of them do, they are less familiar than adult lobsters, crabs, and prawns. I regret that in this book I have been able to show only a small number of them. See the splendid *Atlas of Crustacean Larvae*, published by Johns Hopkins University Press, for the full and amazing range of diversity that these mesmerising little creatures display. Sir Thomas Browne (1605–82) was unaware of them when he wrote the following, about bees, ants, and spiders, but crustacean larvae might have moved him to even greater eloquence.

> Ruder heads stand amazed at those prodigious pieces of nature, Whales, Elephants, Dromedaries and Camels; these I confess, are the Colossus and Majestick pieces of her hand but in these narrow Engines there is more curious Mathematicks, and the civilitie of these little Citizens more neatly sets forth the wisdome of their Maker.

7

In Living Memory

The most recent scripts, those in the top layer of the palimpsest, are those written during the animal's own lifetime. I said that the genes inherited from the past can be seen as predicting the world into which an animal is going to be born. But genes can predict only in a general way. Conditions change on a timescale faster than the generational turnover with which natural selection can cope. Many details are usefully filled in during the animal's own lifetime, mostly by memories stored in the brain, as opposed to the genetic book of the dead, in which 'memories' are written in DNA. Like gene pools, brains store information about the animal's world, information that can be used to predict the future, and hence aid survival in that world. But brains can do it on a swifter timescale. Strictly speaking, where learning – indeed, this whole chapter – is concerned, we are talking not about the genetic book of the dead but about the non-genetic book of the living. However, as we shall see, naturally selected genes from the past prime the brain to learn certain things rather than others.

The gene pool of a species is sculpted by the chisels of natural selection, with the result that an individual, programmed as it is by a sample of genes drawn from the well-carved gene pool, tends to be good at surviving in environments that did the carving: that is, an averaged set of ancestral environments. An important part of the body's equipment for survival is the brain. The brain – its lobes and crevices, its white matter and grey matter, its bewildering byways of nerve cells and highways of nerve trunks – is itself sculpted by natural selection of ancestral genes. The brain is subsequently changed further by learning, during the animal's lifetime, in such a way as to improve yet further the animal's survival. 'Sculpting' might not seem so appropriate a word here. But the analogy between learning and natural selection has impressed many, not least BF Skinner, a leading – if controversial – authority on the learning process.

Skinner specialised in the kind of learning called *operant conditioning*, using a training apparatus that later became known as the Skinner Box. It's a cage with an electrically operated food dispenser. An animal, often a rat or a pigeon, gets used to the idea that food sometimes appears in the automatic dispenser. Built into the wall of the box is a pressable lever or a peckable key. Pressing the lever or key causes food to be delivered, not every time but on some automatically scheduled fraction of occasions. Animals learn to operate the device to their advantage. Skinner and his associates have developed an elaborate science of so-called operant conditioning or reinforcement learning. Skinner Boxes have been adapted to a wide variety of animals. I once saw a film of a rotund gourmand, in a specially reinforced Skinner Box, noisily exercising the lever-bashing skill of his bulbous pink snout. I found it endearing, and I hope the pig enjoyed it as much as I enjoyed the spectacle.

You can train an animal to do almost anything you like, by operant conditioning, and you don't have to use the automated Skinner Box apparatus. Suppose you want to train your dog to 'shake hands', that is, politely raise his right front paw as if to be shaken. Skinner called the following technique 'shaping'. You watch the animal, waiting until he spontaneously makes a move that you perceive as being slightly in the right direction: an incipient, tentative, upward movement of the right front paw, say. You then reward him with food. Or perhaps not with food but with a signal such as the sound of a 'clicker', which he has previously been taught to associate with a food reward.

The clicker is known as a secondary reward or secondary reinforcement, where the food is the primary reward (primary reinforcement). You then wait until he moves his right front paw a little further in the right direction. Progressively, you 'shape' his behaviour closer and closer to the target you have chosen, in this case 'shaking hands'. You can use the same shaping technique to teach a dog to do all manner of cute tricks, even useful ones like shutting the door when there's a cold draught and you are too lazy to get out of your armchair. It is elaborations of the same shaping technique that erstwhile circus trainers employed to teach bears and lions to do undignified tricks.

I think you can see the analogy between behaviour 'shaping' and Darwinian selection, the parallel that so appealed to Skinner and many others. Behaviour-shaping by reward and punishment is the equivalent of shaping the bodies of pedigree dogs by artificial selection – domestic breeding. The gene pools of pedigree cattle, sheep, and cats, of racehorses and greyhounds, pigs and pigeons, have been carefully sculpted by human breeders over many generations to improve running speed, milk or wool yield, or in the case of dogs, cats, and pigeons, aesthetic appeal according to various more-or-less bizarre standards. Darwin himself was an enthusiast of the pigeon fancy, and he devoted an early chapter of *On the Origin of Species* to the power of artificial selection to modify domestic animals and plants.

Now, back to shaping in Skinner's sense. The animal trainer has a particular end result in mind, such as handshaking in a dog. She waits for spontaneous 'mutations' (please note well the quotation marks) of behaviour thrown by an individual animal and *selects* which ones to reward. As a consequence of the reward, the chosen spontaneous variant is then 'reproduced' by the animal itself in the form of a repetition. Next, the trainer waits for a new 'mutant' (again please don't ignore the quotation marks) extension of the desired behaviour. When the dog spontaneously goes a little further in the desired direction of the handshake, she rewards him again. And so on. By a careful regimen of selective rewards, the trainer shapes the dog's behaviour progressively towards a desired end.

The analogy with genetic selection is evident and was expounded by Skinner himself. But so far, the analogy is with artificial selection. How about natural selection? What role does reinforcement learning play in the wild, where there are no human trainers? Does the analogy with reward learning extend from artificial selection to natural selection. How does reward learning improve the animal's survival?

Darwin bridged the gap from domestic breeding to natural selection with his great insight that human breeders aren't necessary. Human selective breeders – let's call them gene pool sculptors – are replaced by natural sculptors: the survival of the fittest, differential survival in wild environments, differential success in attracting mates and vanquishing sexual rivals, differential parenting skills, differential success in passing on genes. And just as Darwin showed that we don't need a human breeder, the analogy with learning does without a human trainer. With no human trainers, animals in the wild learn what's good for them and shape their behaviour so as to improve their chances of survival.

'Mutation' consists of spontaneous *trial* actions that might be subject to 'selection' – i.e. reward or punishment. The rewards and punishments are doled out by nature's own trainers. When a hen scratches the ground with her feet, the action has a good chance of uncovering food of some kind, perhaps a grub or a seed. And so ground-scratching is rewarded, and repeated. When a squirrel bites the kernel of a nut, it's hard to crack unless held at a particular angle in the teeth. When the squirrel spontaneously discovers the right angle of attack, the nut cracks open, the squirrel is rewarded, the correct alignment of the teeth on the nut is remembered and repeated, and the next nut is cracked more quickly.

Much depends on the rewards that nature doles out. Food is not the only reward that we can use, even in the lab. Once, for a research project that I needn't go into, I wanted to train baby chickens to peck differently coloured keys in a Skinner Box. There were reasons not to use food as reward, so I used heat instead. The reward was a two-second blast from a heat lamp, which the chicks found agreeable, and they readily learned to peck keys for the heat reward. But now we need to face the question, what, in general, do we mean by 'reward'? As Darwinians, we must expect that natural selection of genes is ultimately responsible for determining what an animal treats as rewarding. It's not obvious what will be rewarding, however obvious it might seem to us because we are animals ourselves.

We may define a reward as follows. If a random act by an animal is reliably followed by a particular sensation and if, in consequence, the animal tends to repeat the random act, then we recognise that sensation (presence of food or warmth or whatever it is) as a *reward* by definition. If a Skinner Box delivered not food or heat but an attractive and receptive member of the opposite sex, I have no doubt that it would – at least under some circumstances – fit the definition of a reward: an animal in the right hormonal condition would learn to press a key to obtain such a reward. A mother animal, cruelly deprived of her child, would learn to press a key to restore access. And the child would learn to press a key to obtain access to its lost mother. I know of no direct evidence for any of those guesses, nor for my conjecture that a beaver would treat access to branches, stones, and mud suitable for dam-building as a reward by the above definition. And a crow in the nesting season would define access to twigs as a reward. But as a Darwinian, in all those cases I make the prediction with a modicum of confidence.

Brain scientists are able to implant electrodes painlessly in the brains of animals, through which they can stimulate the brain electrically. Normally they do this in order to investigate which parts of the brain control which behaviour patterns. The experimenter controls an animal's behaviour by passing weak electric currents. Stimulate a chicken's brain here, and the bird shows aggressive behaviour. Stimulate a rat's brain there, and the rat lifts its right front paw. The neurologists James Olds and Peter Milner conceived a variant of the technique. They handed the switch over to the rat. By pressing a lever, rats were able to stimulate their own brain. Olds and Milner discovered particular areas of the brain where self-stimulation by rats was highly rewarding: the rats appeared to become addicted to lever-pressing. Not only did electrical stimulation in these brain regions fulfil

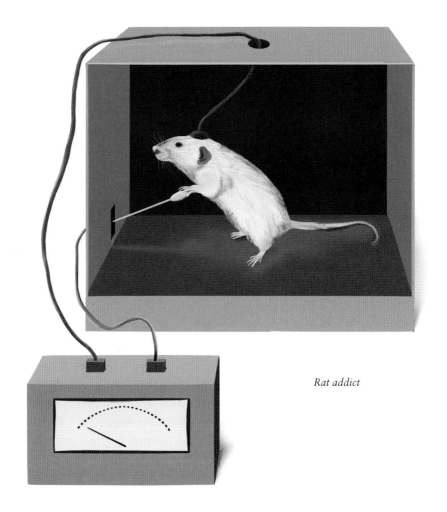

Rat addict

the definition of a reward. It did so in a big way. When the electrodes were inserted in these so-called pleasure centres, rats would obsessively press the switch, to the extent of unfortunately neglecting other vital activities. They would sometimes press the lever at a rate of 7,000 presses per hour, would ignore food and receptive members of the opposite sex and go for the lever instead, would run across a grid delivering electric shocks in order to get at the lever. They would press the lever continually for twenty-four hours until the experimenters removed them for fear they'd die of starvation. The experiments have been repeated on humans with similar results. The difference is that humans could verbalise what it felt like:

A sudden feeling of great, great calm … like when it's been winter, and you have just had enough of the cold, and you go outside and discover the first little shoots and know that spring is finally coming.

Another woman (and you have to wonder whether the experiment was approved by an ethics committee)

quickly discovered that there was something erotic about the stimulation, and it turned out that it was really good when she turned it up almost to full power and continued to push on her little button again and again … she often ignored personal needs and hygiene in favor of whole days spent on electrical self-stimulation.

It seems plausible that natural selection has wired up animal brains in such a way that external stimuli or situations that are good for the animal (which will vary from species to species) are internally connected to the 'pleasure centres' discovered by Olds and Milner.

Punishment is the opposite of reward. If an action is reliably followed by a stimulus X and, as a consequence, the animal becomes less likely to repeat the action, then X is defined as a punishment. In the laboratory, psychologists sometimes use electric shock as punishment. More humanely (I guess) they use a 'time out' – an interval during which the animal is denied access to reward. Dog trainers (the practice is frowned upon by many experts, rightly in my opinion) sometimes smack an animal as punishment. When I was at boarding school (and this practice is now not only frowned upon but illegal) my friends and I were from time to time caned by the headmaster, hard enough (astonishing as it now seems) to leave bruises that took weeks to heal (and were admired at bath-time like battle scars). What my offences were I have now forgotten, but I'm sure I didn't forget while I was still at the school and within range of Slim Jim and Big Ben, the two canes in the headmaster's quiver. My probability of repeating the offence undoubtedly decreased. Therefore, beatings were punishments by definition, as well as by the intention of the headmaster.

In nature, bodily injury is perceived as painful. If an action is followed by pain, the probability of repeating that action goes down. Not only is that how we define punishment: it also explains what pain is for, in the Darwinian sense. Injury often presages death and hence failure to reproduce. Therefore, the nervous system defines bodily injury as painful.

Sometimes pain is endured when offset by reward. We've already seen that rats will endure painful electric shock to get to the self-stimulation lever. The punishment of a bee sting may be offset by the reward of honey. The taste of honey is such an intense reward that many animals, including bears, honey badgers, raccoons, and human hunter-gatherers, are prepared to endure the pain for the sake of it. Rewards and punishments trade off against each other, just as mutually opposing natural selection pressures trade off against each other.

The Darwinian interpretation of pain as a warning not to repeat the preceding action has ethical implications. In our treatment of non-human animals, on farms and hunting fields, in slaughterhouses and bullrings, we are apt to assume that their capacity to suffer is less than ours. Are they not less intelligent than we are? Surely this means they feel pain, if at all, less acutely than us? But why should we assume that? Pain is not the kind of thing you need intelligence to experience.

The capacity to feel pain has been built into nervous systems as a warning, an aid to learning not to repeat actions that caused bodily damage and might next time lead to death. So, if a species is less intelligent, might its pain need to be more agonising, rather than less? Shouldn't humans, being cleverer, get away with less painful pain in order to learn not to repeat the self-harming action? A clever animal, you might think, could get away with a mild warning, 'Er, probably a good idea not to do that again, don't you think?' Whereas a less intelligent animal would need the sort of dire warning that only excruciating pain can deliver. How should this affect our attitude towards slaughterhouses and agricultural husbandry? Should we not, at very least, give our animal victims the benefit of the doubt? It's a thought, to put it at its mildest!

Rewards and punishments, pleasure and pain, are so familiar and obvious to us as human animals that you probably wonder why I am labouring the topic in this chapter. Here is where things start to become less obvious and more interesting. The brain's choice of what shall constitute reward and what punishment is not fixed in stone. It is ultimately determined by genetic natural selection. Animals come into the world equipped with genetically granted definitions of reward and punishment. These definitions have been made by natural selection of ancestral genes. Any sensation associated with an increased probability of death will become defined as painful. A dislocated limb in the wild dramatically increases the probability of death. And it is intensely painful, as I recently and very vocally testified, all the way to the hospital. It has certainly made me take great care to avoid risking a repeat. Copulation increases the probability of reproduction, and genetic selection has consequently made the accompanying sensations pleasurable – which means rewarding. It has been suggested, with support from rat experiments and from the self-stimulating woman mentioned above, that sexual pleasure is directly linked to the 'pleasure centres' discovered by Olds and his colleagues. Presumably other sensations, too, could be so linked by natural selection.

I conjecture that by artificial selection you could breed a race of pigeons who enjoy listening to Mozart but dislike Stravinsky. And vice versa. After many generations of selective breeding, perhaps spread over several human lifetimes, the birds would be genetically equipped with a definition of reward such that they would learn to peck a key that caused a recording of Mozart to be played, and would learn to peck a key that caused a recording of Stravinsky to be switched off. And of course, the experiment would be incomplete unless we also bred a line of pigeons who treated Mozart as punishment and

Stravinsky as reward. Let's not get pedantic as to whether it is really Mozart that they'd treat as rewarding. The learned preference would probably generalise from Mozart to Haydn! The only point I am trying to make is that the definitions of what is rewarding and what is punishing are not carved in stone. They are carved in the gene pool and therefore potentially changeable by selection.

As a corollary, I conjecture that, by artificial selection, you could (though I wouldn't wish to, and it might take an unconscionable number of generations) breed a race of animals who regarded what had previously been pain as rewarding. By definition, it would no longer be pain! It would be cruel to release them into their species' natural environment because, of course, they would be unfitted to survive there – that's the whole point. But the mere fact that they enjoy what normal members of their species would call pain is not cruel – because, however hard it is for us to imagine, at least within the confines of my thought experiment, they enjoy it! Anyway, the more interesting conclusion is that, in a state of nature, it is natural selection that determines what is reward and what is punishment. My thought experiment was devised to dramatise that conclusion.

Experimental psychologists have long known that you can *train* an animal to treat as a reward something that previously had neutral value for the animal. As mentioned above, it's called *secondary reinforcement*, and an example is the clicker used by dog trainers. But secondary reinforcement is not what I'm talking about here, and I really want to emphasise that. I'm not talking about secondary reinforcement, but about genetically changing the very definition of what constitutes primary reinforcement. I conjecture that we could achieve it by breeding, as opposed to training. I called it a conjecture because the experiment has not, as far as I know, been done. I'm now talking about selectively breeding animals in such a way as to change their own genetically instilled definition of what constitutes a primary reward in training. To repeat my suggestion above, I predict that by artificial selection you could in principle breed a race of animals who would treat bodily injury as rewarding.

Douglas Adams carried the point to a wonderful comedic *reductio* in *The Restaurant at the End of the Universe*. Zaphod Beeblebrox's table was approached by a large bovine creature, who announced himself as the dish of the day. He explained that the ethical problem of eating animals had been solved by breeding a species that wanted to be eaten and was capable of saying so. 'Something off the shoulder, perhaps, braised in a white wine sauce?'

Birds don't naturally listen to human music, so my Mozart/Stravinsky flight of fancy may seem implausible. But do they have a music of their own? A respected ornithologist and philosopher named Charles Hartshorne suggested that we should regard bird-song as music, appreciated aesthetically by the birds themselves. He may not have been wrong, as I shall soon argue.

The role of learning and genes in the development of birdsong has been intensively studied, especially by WH Thorpe, Peter Marler, and their colleagues and students. Many birds learn to imitate the song of their father or other members of their own species. Spectacular feats of mimicry by the likes of mynahs and lyre birds are an extreme. In addition to mimicking other species such as kookaburras ('laughing jackass'), lyre birds have been recorded by David Attenborough giving remarkably convincing imitations of car alarms, camera shutters (with or without a motor drive), the chainsaws of lumberjacks and the mixed noises of a building site. I have even heard it said, but have failed to verify it, that lyre birds can distinctly mimic Nikon versus Canon camera shutters. Such virtuoso mimics incorporate an amazing variety of such sounds in an ample repertoire.

This raises the question of why many songbirds have large repertoires in the first place. Individual male nightingales can sport more than 150 recognisably distinct songs. Admittedly that's an extreme, but the general phenomenon of song repertoires demands an explanation. Given that song serves to deter rivals and attract mates, why not stick to one song? Why switch between alternatives? Several hypotheses have been proposed. I'll mention just my favourite, the 'Beau Geste' hypothesis of John Krebs.

In the adventure yarn of that name by PC Wren, an outnumbered unit of the French Foreign Legion was beleaguered in a desert fort,

and the commander beat off the opposing force with a spectacular bluff.

> As each man fell, throughout that long and awful day, [the com-
> mander] had propped him up, wounded or dead, set the rifle
> in its place, fired it, and bluffed the Arabs that every wall and
> every embrasure and loophole of every wall was fully manned.

Krebs's hypothesis is that the bird with a large repertoire is pretending
his territory is already occupied to the full. He is, as it were, mimicking the sounds that would emerge from an area if it were already
overpopulated with too many members of his species. This deters
rivals from attempting to set up their territory in the area. The more
densely populated an area is, the less will it benefit an individual
to settle there. Above a certain critical density, it pays an individual to leave and seek territory elsewhere, even an otherwise inferior
territory. So, by pretending to be many nightingales, an individual
nightingale seeks to persuade others to find a different place to set
up his territory. In the case of lyre birds, the sound of a chainsaw is
just another addition to the repertoire, the size of which conveys the
message: 'Go away, there's no future for you here, the place is fully
occupied.'

Virtuoso impressionists like lyre birds, mynahs, parrots, and
starlings are outliers. Probably they are just manifesting, in extreme
form, the normal way young birds learn their species song – imitating their fathers or other species members. The point of learning the
correct species song is to attract mates and intimidate rivals. And
now we return to our discussion of the definition of a reward: how
natural selection defines what will be treated as reward and what
punishment.

In an experiment by JA Mulligan, three American song sparrows
(*Melospiza melodia*) were reared by canaries in a soundproof room
so that they never heard the song of a song sparrow. When they grew
up, all three produced songs that were indistinguishable from those
of typical wild song sparrows. This shows that song sparrow song

is coded in the genes. But it is also learned. In the following special sense. Young song sparrows teach themselves to sing, with reference to a built-in template, a genetically installed idea of what their song ought to sound like.

What's the evidence for this? It is possible surgically, under anaesthetic and I trust painlessly, to deafen birds. This has been done, with both song sparrows and the related white-crowned sparrows, *Zonotrichia leucophrys*. If birds of either species are deafened as adults, they continue to sing almost normally: they don't need to hear themselves sing. As adults, that is. If, however, they are deafened when three months old, too young to sing, their song when they reach adulthood is a mess, bearing little resemblance to the correct song. On the template hypothesis, this is because they have to teach themselves to sing, matching their random efforts against the template of correct song for the species. There's an interesting difference between the two species. Whereas the song sparrow never needs to hear another bird sing – its template is innate – the white-crowned sparrow makes a 'recording' of white-crowned sparrow song, early in life, long before it starts to develop its own song. Once the template is in place, whether innate as in the song sparrow or recorded as in the white-crowned, the nestlings then use it to teach themselves to sing.

Doves and chickens push this to an extreme: they don't need to listen to themselves, ever. Ring dove (also known as barbary dove) squabs, who have been surgically rendered completely deaf, later develop vocalisations that are just like those of intact doves. That the behaviour is innate is further testified by the fact that hybrid doves coo in a way that is intermediate between the parental species' coos. As we shall see in Chapter 9, young crickets (nymphs), before they achieve their final moult to become adults, can artificially be induced to display nerve-firing patterns identical to their species song patterns, even though nymphs never sing. And hybrid crickets have a song that is intermediate between the two parental species.

But I want to get back to the sparrows. As we have seen, they teach themselves to sing by listening to their own random babblings, and repeating those fragments that are rewarded by a match to a template – whether the template is genetically built-in (song sparrow), or a 'recording' (white-crowned sparrow) remembered from infancy. Did you notice this means that a sound that matches the template is a reward by our definition? We have identified a new kind of reward to add to food and warmth. The song template is a much more specialised kind of reward. It's easy to see how food (relief of hunger pangs) and warmth (relief of cold discomfort) would be general, non-specific rewards. Indeed, psychologists of the early twentieth century delighted in reducing all rewards to one simple formula, which they called 'drive reduction'. Hunger and thirst were seen as examples of 'drives', analogous to forces *driving* the animal. A particular pattern of sounds, complicated and characteristic enough to be recognised, by ornithologists and birds alike, as belonging to one species and one species alone, is a reward of a very different kind from generalised drive-reduction. And, I would personally add, of a much more interesting kind. As a student I tried to read up that rat psychology literature, and I'm sorry to admit that I found it rather boring compared to the zoology literature on wild animals.

The ethologist Keith Nelson once gave a conference talk with the title 'Is bird song music? Well, then, is it language? Well, then, what the hell is it?' It isn't language: not rich enough in information, and it doesn't seem to be grammatical in the sense of possessing a hierarchical nesting of 'clauses' enclosing 'sub-clauses'. Hartshorne, as I mentioned previously, thought it was music, and I think there's a sense in which he was right. I believe we can make a case that birds have an aesthetic sense, which responds to song. I think there's also a sense in which it works like a drug. In what follows, I am drawing on a pair of papers that I wrote jointly with John Krebs some years ago, about animal signals generally. We were critically responding to a then prevalent idea that animal signals function to convey useful *information* from the sender to the recipient, for the mutual benefit of both. For example, 'I am a male of the species *Luscinia megarhynchos*, I am in breeding condition, and I have a territory over here.' The gene's-eye view of evolution, then quite novel, did not sit well with 'mutual benefit'. Krebs and I followed the gene's-eye view to a more cynical view of animal signals, substituting the idea of *manipulation* of the receiver by the signaller. 'You are a female of the species *Luscinia megarhynchos*. COME HITHER! COME HITHER! COME HITHER!'

> When an animal seeks to manipulate an inanimate object, it has only one recourse – physical power … But when the object it seeks to manipulate is itself another live animal there is an alternative way. It can exploit the senses and muscles of the animal it is trying to control … A male cricket does not physically roll a female along the ground and into his burrow. He sits and sings, and the female comes to him under her own power.

Now, you might object, surely the female should respond to male song in this way only if it benefits her. But we regarded the relationship between signaller and signallee as an arms race, run in evolutionary time. Perhaps she does put up some sales-resistance. But that provokes the male, on the other side of the arms race, to up the ante: increase the intensity of his signal. And now we come to another strand to the argument, which Krebs and I advanced in the second of our two papers. This concerns what we called 'mind-reading'. Any animal in a social encounter can benefit itself by predicting (behaving as if predicting) the behaviour of another. There are all kinds of give-away clues. If a male dog raises his hackles, this is an involuntary indicator of an aggressive mood. Responding appropriately to such give-aways is what we dubbed 'mind-reading'. Humans can become quite adept at mind-reading in this sense, making use of such cues as shifty eyes or fidgety fingers. And now, to bring the argument full circle, an animal who is the victim of a mind-reader can exploit the fact of being mind-read, in such a way as to render inappropriate the very word 'victim'. A male, for instance, might manipulate a female by 'feeding' her mind-reading machinery, perhaps with deceptive cues. This is just to say that where victimhood is concerned, manipulation is not a one-way street. Mind-reading turns the tables. And then manipulation potentially turns them back again, against the mind-reader.

On this view animal signals, to repeat, evolve as an arms race between mind-reading and manipulation, an arms race between salesmanship and sales-resistance. In those cases where the sender benefits from being mind-read and the receiver benefits from being manipulated, we suggested that the ensuing signal should shrink to a 'conspiratorial whisper'. Why escalate a signal when there is no push-back. Conversely – the opposite of a conspiratorial whisper – loud, conspicuous, vivid signals will arise where the recipient does not 'want' to be manipulated. In such cases the arms race, in evolutionary time, escalates towards exaggeration on the part of the sender, to combat increased 'sales-resistance' on the part of the receiver.

Why, you might wonder, should there ever be 'sales-resistance'? It's most easily seen in the case of the arms race between the sexes. You might think it's always a good idea for males and females to get together and coordinate. You'd be wrong, and for an interesting reason. Ultimately because sperms are smaller and more numerous ('cheaper') than eggs, females need to be choosier than males. A male is more likely to 'want' to mate with a female than the female will 'want' to mate with him. Females pay a higher cost if they mate with the wrong male than males pay if they mate with the wrong female. In extreme cases, there is no such thing as the wrong female. Hence males are more likely to escalate salesmanship when trying to persuade females. And females more likely to favour sales-resistance. Where you see high-amplitude signals – bright colours, loud sounds – that means there's probably sales-resistance. Where there's no sales-resistance, signals are likely to sink to a conspiratorial whisper. Conspicuous signals are costly, if not in energy, in risk of attracting predators or alerting prey.

I've been a bit terse in condensing two full-sized papers into four paragraphs. It should become clearer when I now apply it to birdsong. Birdsong is too loud and conspicuous to be a 'conspiratorial whisper', so let's go for the other extreme: increased sales-resistance fomenting exaggerated efforts to manipulate. Is birdsong an attempt to manipulate the behaviour of females and other males: an attempt to change their behaviour to the advantage of the singer?

If biologists wish to manipulate the behaviour of a bird, what can they do? This chapter has already introduced one possibility that birds themselves, unfortunately for them, cannot do: electrical stimulation of another's brain through implanted electrodes. The Canadian surgeon Wilder Penfield pioneered the technique on human patients whose brains were undergoing surgery for other reasons. By exploring different parts of the cerebral cortex, he was able to jerk specific muscles into action like a puppeteer pulling strings. When he drew

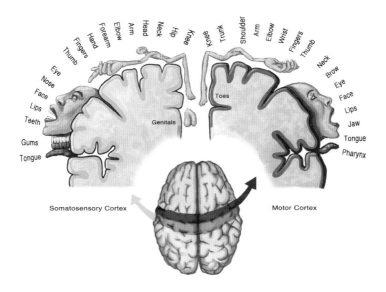

a map of which parts of the brain pulled which muscles, it looked like a caricature of a human body, the so-called 'motor homunculus' (there's also a 'sensory homunculus' on the left-hand side of the picture, which looks rather similar). The grotesque exaggeration of the homunculus's hand goes some way towards explaining the formidable skill of a concert pianist, for example. And the large brain area given over to the lips and tongue is no doubt related to speech. The German biologist Erich von Holst, working with chickens in a deeper part of the brain, the brain stem, was able to control what might be called the bird's 'mood' or 'motivation', resulting in changes to the observed behaviour, including 'guiding hen to nest' and 'uttering call warning of predator'. I repeat that these operations are painless, by the way. There are no pain receptor nerves in the brain.

Now, a male nightingale might well 'wish' he could implant electrodes in a female's brain and control her behaviour like a puppet. He can't do that, he's no von Holst, and he has no electrodes. But he can sing. Might song have something like the same manipulative effect? No doubt he might benefit, if only he could inject hormones into her bloodstream. Again, he can't literally do that. But evidence on ring doves and canaries suggests that birds can do something close to it. Male doves vigorously court females with a display called the bow-coo. The bow is a characteristic movement resembling an unusually obsequious human bow, and it is accompanied by an equally characteristic coo, consisting of a staccato note followed by a purring *glissando*. A week's exposure to a bow-cooing male reliably causes massive growth of a female's ovary and oviduct, with accompanying changes in sexual, nest-building, and incubation behaviour. This was shown by the American animal psychologist Daniel S Lehrman. Lehrman went on to show that the behaviour of male ring doves has a direct effect on the hormones circulating in female bloodstreams. Parallel work by Robert Hinde and Elizabeth Steel in Cambridge on nest-building behaviour in female canaries showed the same thing.

The ring dove and canary type experiments have not been done on nightingales, but it probably is generally the case that male birdsong changes the hormonal state of females. Male song manipulates female behaviour, as though the male had the power to inject her with chemicals, presumably nightingales no less than other species.

> *My heart aches, and a drowsy numbness pains*
> *My sense, as though of hemlock I had drunk,*
> *Or emptied some dull opiate to the drains*
> *One minute past, and Lethe-wards had sunk.*

John Keats was not a bird, but his brain was a vertebrate brain like a female nightingale's. The male nightingale song drugged him – almost to death in his poetic fancy. If it can so intoxicate the mammal Keats, might it not have a yet more powerful effect on the vertebrate brain that it was designed to beguile, the brain of another nightingale? To answer yes, we hardly need the testimony of the dove and canary experiments. I believe natural selection has shaped the male nightingale's song, perfecting its narcotic power to manipulate the behaviour of a female, presumably by causing her to secrete hormones.

But now, let's return to learning, and the deafening experiments. The evidence shows that young white-crowned sparrows and song sparrows teach themselves to sing with reference to a template. Young white-crowneds need to hear song in order to make their 'recording' of the template. But any old song won't do. They have to hear the song of their own species. This shows that, even when the template is recorded, there is an innate component to it, built in by the genes. And in the case of the song sparrow, it doesn't even need to be recorded.

I suggested above that birdsong might be appreciated as music, enjoyed aesthetically by the birds themselves. We are now in a position to spell out the argument. The male teaches himself to sing by comparing his 'random' burblings against a template. The template serves as reward, positively reinforcing those random attempts that happen to match it. Reflect, now, that the male songster has a brain much like the female he later hopes to manipulate. When he teaches himself to sing, he is finding out which fragments of song appeal to a bird of his own species (himself ... but later, a female). What is that, if not the employment of aesthetic judgment?

Burble. I like it (conforms to my template). Repeat it.
Burble warble. Ooooh, that's even better. I like that very much.
It really turns me on. Repeat that too. YES!

What turns him on will probably turn a female on too, for they are, after all, members of the same species with the same typical brain of the species. At the end of the developmental period when the final adult song has been perfected, it will be equally beguiling to the singer himself and his female target. He learns to sing whichever phrases turn him on. There seems no powerful reason to deny that both enjoy an aesthetic experience – as did John Keats when he heard the nightingale.

We've come a long way from the idea of reward as generalised 'drive reduction'. And we've arrived at what I think is a much more interesting place. The lesson of these experiments on birdsong is that reward can be a highly specific stimulus, or stimulus-complex, ultimately laid down by genes: what Konrad Lorenz, one of the fathers of ethology, dubbed the 'Innate Schoolmarm'.

If this is right, we should predict the following result in a Skinner Box. A young song sparrow who has never heard song should learn to peck a key that yields the sound of song sparrow, but no other species' song. That hasn't been done, but various similar experiments have. Joan Stevenson found that chaffinches preferred to settle on a perch attached to a switch that turned on chaffinch song. However, the control sound for comparison was white noise, not the song of another species. Her chaffinches, moreover, were not naive but wild caught. Her method was adopted by Braaten and Reynolds with hand-reared, naive zebra finches and using starling song for comparison instead of white noise. They showed a clear preference for perches that played zebra finch song rather than starling song. It would be nice to do a big experiment with, say, naive young songbirds of six different species, with six perches, each perch playing one of the six songs. We should predict that each species should learn to sit on the perch that played their own species song. It wouldn't be an easy experiment. Hand-rearing baby songbirds is hard work. A neat design might be to give each baby to foster parents of one of the other six species.

The template of song sparrows is innate. The 'recorded' template of young white-crowned sparrows, laid down early in life before they start singing, looks like the kind of learning called 'imprinting', most closely associated with Konrad Lorenz and his pursuing geese. Imprinting was first recognised in *nidifugous* baby birds.

Nidifugous, from the Latin, means 'fleeing the nest'. Nidifugous hatchlings start life equipped with warm and protective downy feathers and well-coordinated limbs. Examples are ducklings, goslings, moorhen chicks, chicken chicks, ground-nesting species generally. Within hours of hatching, as soon as their feathers are dry, nidifugous chicks are up and about, walking competently, looking around alertly, pecking at food prospects, and dogging parental footsteps. The opposite of nidifugous is *nidicolous*. All songbirds are nidicolous. Nidicolous bird species typically nest in trees. The babies are helpless, naked, incapable of walking (they're in a nest balanced up a tree, where would they walk to?), incapable of feeding themselves but with a huge gaping beak, a begging organ into which their parents tirelessly shovel food. Many seabirds such as gulls are nidifugous in that they hatch with downy feathers and don't gape for food. But they are dependent on the parents bringing food that they regurgitate for the chicks.

Mammals, too, have their own equivalent to nidifugous (think gambolling lambs; and wildebeest calves must follow the herd on the day they're born) and nidicolous (baby mice are hairless and helpless). Man is a nidicolous species. Our babies are almost completely helpless. There has been an evolutionary trade-off between a pressure towards a bigger brain, conflicting with the difficulty of being born imposed by a large head. The result was to push our babies towards being born earlier, before the head became insufferably (for the mother) large to push out. The result was to make us even more helplessly nidicolous than other ape species.

Nidifugous species, both mammals and birds, are in danger if they become separated from their parent(s), and this is where imprinting comes in. Nidifugous babies, as soon as they hatch, do something equivalent to taking a mental photograph of the first large moving object they see. They then follow it about, at first very closely, then venturing gradually further away as they grow older. The first moving object they see is usually their parent, so the system works fine in nature. Goslings hatched in an incubator, however, tend to imprint on a human carer, for example Konrad Lorenz.

The idea of imprinting in mammals is imprinted in child minds by the nursery rhyme 'Mary had a little lamb' (Everywhere that Mary went / The lamb was sure to go). Imprinted animals, both birds and mammals, often retain their mental photograph into adulthood and attempt to mate with creatures (such as humans) who resemble it. One of the reasons zoos have difficulty with breeding is that the frustrated animals hanker after their keepers.

Imprinting may or may not be a special kind of learning. Some say it's just a special case of ordinary learning. It's controversial. Either way, it's a nice example of a recent, 'top layer' palimpsest script. The genes could have equipped the animal with a built-in image or specification of precisely what to follow, what to mate with, what song to sing. Instead, they equip the animal with rules for colouring in the details.

Reinforcement learning and imprinting are not the only kinds of learning by which an animal, during its own lifetime, supplements the inherited ancestral wisdom. Elephants make important use of traditional knowledge. The brains of old matriarchs contain a wealth of knowledge about such vital matters as where water can be found. Young chimpanzees learn from their elders skills such as using a stone as a hammer to crack nuts, and preparing a twig to probe termite nests. The handover from adept to apprentice is a kind of inheritance, but it is memetic, not genetic. This is why these skills are practised in particular local areas and not others. The skill of sweet potato washing in Japanese macaques is another example. So is pecking through the foil or cardboard lids of milk bottles by British tits, in the days when milk was delivered daily on the doorstep. In this case, the skill was seen to radiate geographically outwards from focal points, in the manner of an epidemic.

What else equips animals to improve on their genetic endowment, apart from learning? Perhaps the most important example of a 'memory' not mediated by the brain is the immune system. Without it, none of us would have survived our first infection. Immunology is a

Geese imprinted on Konrad Lorenz. A special kind of learning,
which casts light on the mind of birds

huge subject, too big for me to do it justice in this book. I'll say a few words, just enough to make the point that genes don't attempt the impossible task of equipping bodies with information about all the bacteria, viruses, and other pathogens that they might ever encounter. Instead, genes furnish us with tools for 'remembering' past infections, forearming us against future infection. We carry not just the genetic book of the dead (the ancestral past) but a special molecular book in which is written a continually updated medical record of our infections and how we dealt with them.

Bacteria, too, suffer from infection – by viruses called bacteriophages, or phages for short – and they have their own immune system, which is rather different from ours. When a bacterium is infected, it stores a copy of part of the viral DNA within its own single circular chromosome. These copies have been called 'mug shots' of criminal viruses. Each bacterium sets aside a portion of its circular chromosome as a kind of library of these mug shots. The mug shots will later be used to apprehend criminals in the form of the same or related viruses making a reappearance. The bacterium makes RNA copies of the mug shots. These RNA images of 'criminal' DNA are circulated through the interior of the bacterial cell. If a virus of a familiar type should invade, the appropriate mug shot RNA binds to it, and special protein enzymes cut up the joined pair, rendering the virus harmless.

The bacterium needs a way to label the mug shots, so they aren't confused with its own DNA. They are labelled by the presence of adjacent nonsense sequences of DNA, which are palindromes called CRISPR: Clustered Regularly Interspaced Short Palindromic Repeats. Each time a bacterium is assailed by a new kind of virus, another CRISPR-flanked mug shot is added to the CRISPR region of the chromosome. It's another story, but CRISPR has become famous because scientists have discovered a way in which the bacterial skill can be borrowed for the human purpose of editing genomes.

The vertebrate immune system works rather differently. It's more complicated but we too have a 'memory' of pathogens of the past. Our immune system is then able to mount a rapid response, should any of those old enemies venture to return. This is why those of us who have had mumps or measles can safely mingle with victims, confident that we shall not get the disease a second time. And the enormous boon of vaccination works by tricking the immune system into building up a false memory, normally by injecting either a killed strain or a weakened strain of the pathogen.

The Covid-19 pandemic was largely stopped in its tracks, saving thousands of lives, by a wonderful new type of vaccine, the mRNA vaccine. The role of mRNA (messenger RNA) is to convey coded messages from DNA in the nucleus to where proteins are made to the code's specification. Now, here's how mRNA vaccines work. Instead of injecting a killed or weakened strain of the dangerous virus, a harmless protein in its jacket is first sequenced. The genetic code appropriate to that protein is then written into mRNA. The mRNA does its thing, which is to code the synthesis of protein – in this case the harmless jacket protein of the Covid virus. And then, the immune system does *its* thing and attacks the virus if it enters the body, recognising it by the protein in its jacket.

What is especially interesting, in pursuit of our analogy between learning and evolution, is that the vertebrate immune system's 'memory' (unlike the bacterial one) works in a kind of Darwinian way, by an internal version of natural selection, within the body. But that is another story, beyond our scope here.

The immune system, and the brain, are the two rich data banks in which entries are written during the animal's own lifetime, to update the genetic book of the dead, or 'colour in the details'. More minor examples need mentioning for the sake of completeness. Darkening of the skin is a kind of memory of lying out in the sun. It provides useful screening against the damage that the sun's rays, especially ultraviolet, can wreak, for example in causing skin cancers. This is a case where genetic and post-genetic scripts both contribute. People whose ancestors have lived many generations in fierce tropical sun tend to be born with dark skin, for example native Australians, many Africans, and people from the south of the Indian sub-continent. By contrast, those whose ancestors have lived many generations at higher latitudes are at risk from too little sun. They tend to lack Vitamin D and hence are prone to rickets. Genetic natural selection at high latitudes has therefore favoured lighter skins. That's all written in the genetic book of the dead. But this chapter is about palimpsest scripts written after birth, and here is where suntan comes in. Browning in the sun, a post-birth 'colouring-in', achieves in light-skinned, high-latitude people a temporary approach towards what is written into the genome of tropical peoples. You could think of the two as short-term memory and long-term memory of sunlight.

Another example is acclimatisation to high altitude. The higher you go, the thinner the atmosphere, where lack of oxygen causes 'mountain sickness', whose symptoms include headaches, dizziness, nausea, and complications of pregnancy. People whose ancestors have long lived at high altitude have evolved genetic adaptations such as elevated haemoglobin levels in the blood. Those 'memories' of ancestral natural selection are written in the genetic book of the dead. Interestingly, the details differ between Andean and Himalayan peoples, not surprisingly because they have independently, over 10,000 years or more, adapted to a lack of oxygen in mountainous regions widely separated from each other. There are several routes to acclimatisation, and it is not surprising that different mountain peoples have followed different evolutionary paths.

Once again, ancestral scripts can be over-written during the animal's own lifetime. Lowland people who move to high areas can acclimatise. In 1968, when the Olympic Games were held in Mexico City, national teams deliberately arrived early, in order to train at the high altitude (2,200 metres, more than 7,000 feet) of the Anahuac Plateau. Changes that develop during a period of weeks living at high altitude are written into the post-birth palimpsest layer. As with skin colour, they mimic the older, gene-authored scripts.

Talking of skin colour, the 'paintings' of Chapter 2 were all done by ancestral genes, replaying ancestral worlds. But there are some animals who can repaint their skin on the fly, to match the changing background they happen to be sitting on at any given moment. This is another example of the non-genetic book of the living. Chameleons are proverbial, but they aren't the top virtuosi when it comes to

impromptu skin artistry. Flatfish such as plaice can change not just their colour but also their patterning. The one above is capable of changing its colour to match the yellow background on which it now sits. But you only have to take one look at it to read it as a detailed description of the lighter bottom it has just moved off, with its mottled pattern projected by shimmering light from surface ripples.

Even flatfish are upstaged by octopuses and other cephalopod molluscs, who have perfected the art of dynamic cross-dressing to an astonishing extent. And they, uniquely in the animal kingdom, do their changes at high speed. Roger Hanlon, while diving off Grand Cayman, saw a clump of brown seaweed suddenly turn ghostly white and swim rapidly away in a puff of sepia smoke. It was an octopus, with a perfect painting of brown seaweed all over its skin. As Hanlon approached, an emergency order from the octopus brain twitched the muscles controlling the tiny bags of pigment peppering the skin. Instantaneously, the whole surface changed colour from perfect camouflage (trying not to be noticed by predators) to scary white (startling

Thaumoctopus mimicus

Thaumoctopus mimicus

Sea snake

Flounder

would-be predators). Finally, the puff of dark brown ink deflects the attention of would-be predators away from the fleeing octopus.

Hanlon saw an octopus (upper right) in Indonesian waters, *Thaumoctopus mimicus*, who mimicked a flounder (lower right), not just its appearance but also its behaviour, stopping and starting in jerky glides over the sand surface. What's the point? Hanlon is unsure, but he suspects it deceives predators who like to bite off a tentacle but cannot cope with a substantial flatfish. This octopus also can put on a show with its tentacles (upper left), making each one resemble a venomous sea snake (lower left) common in tropical waters. Cephalopods can even change their skin's texture, ruffling up or puckering it into extraordinary shapes. A colleague once dramatised their other-world strangeness by beginning a lecture on Cephalopods: 'These are the Martians.'

The main thesis of this book is that the animal can be read as a description of much older, ancestral environments. This chapter has shown how further details are added, on top of the ancestral palimpsest scripts. Earlier chapters invoked a future scientist, SOF, presented with an animal and challenged to read its body and reconstruct the environments that shaped it. There, we spoke only of ancestral environments, described in the genomic database and its phenotypic manifestations. In this chapter we've seen how SOF could supplement her reading of ancestral environments, by additional readings of the more recent past, including the other two great databases that supplement the genes, namely the brain and the immune system. Today's doctors can read your immune system database and reconstruct a moderately complete history of the infections you have suffered – or been vaccinated against. And if SOF could read what is written in the brain (a big if, she really would have to be a scientist of the future), she could reconstruct much detail of the animal's past environments in its own lifetime.

Experience, either literal experience stored in the brain as memories, disease experience, or genetic 'experience' sculpted into the genome by natural selection, enables an animal to predict (behave as if predicting) what will happen next. But there's one more trick that the brain can pull off in order to foretell the future: simulation, or imagination. Human imagination is a much grander affair than this but, from the point of view of an animal's survival, and our analogy between natural selection and learning, we could regard imagination as a kind of 'vicarious trial and error'. Unfortunately, that particular phrase has been usurped by rat psychologists. A rat in a 'maze' (usually just a choice between turning left or right) will sometimes physically vacillate, looking left, right, left, right before finally making up its mind. This 'VTE' may be a special case of imagining alternative futures, but it's probably safest if I reluctantly surrender the phrase itself to the rat-runners and not use it here. Instead, I'll prefer an analogy with computer simulation: the animal's brain simulates likely consequences of alternative actions internally, thereby sparing itself the dangers of trying them out externally in the real world.

I said the human imagination is a much grander affair. It finds expression in art and literature. Words written by one person can call up an imagined scene in the brain of another. Gertrude's lament for Ophelia can move a reader to tears four centuries after the poet's death. Less ambitiously, let me ask you to imagine a baboon atop a steep cliff. Someone has balanced a plank over the edge of the cliff. Resting at the far end of the plank, over the abyss, is a bunch of bananas. Imagine them, yellow and tempting. The baboon is indeed tempted to venture out along the plank. However, his brain internally simulates the consequence, sees that his extra weight would topple the plank – imagines himself tumbling to his death. So he refrains.

Let's now imagine a range of brains faced with the banana on the plank. First, the genetic book of the dead can build in an innate fear of heights. I myself experience a tingling of the spine, which inhibits me from walking within a metre of the edge of a precipice such as the Cliffs of Moher in Western Ireland. This, even when there's no wind and no reason to suppose that I would fall.

A whole genre of experimentation, the so-called 'visual cliff' experiment, has been devised to investigate fear of heights. The baby in the picture is quite safe: there's strong glass over the 'cliff'. I recently

The visual cliff

visited one of the world's tallest buildings where one could stand on toughened glass looking down on the street far below. Perfectly safe, and I watched others walk on the glass, but I avoided doing so myself. Irrational, but innate fears are hard to conquer. Perhaps an innate fear of heights is inherited from tree-climbing ancestors who survived because they possessed it. Not everyone succumbs, of course. These New York construction workers are enjoying a relaxed lunch with evident (though incomprehensible to me) nonchalance.

Death by falling is the crudest route through which a fear of heights might be built into animals. Another way is by learning, reinforced by pain. Young baboons who fall down smaller cliffs are not killed, but they experience pain. Pain, as we've seen, is a warning: 'Don't do that again. Next time the cliff might be higher, and it will kill you.' Pain is a kind of vicarious, relatively safe substitute for death. Pain stands in for death in the analogy between learning and natural selection.

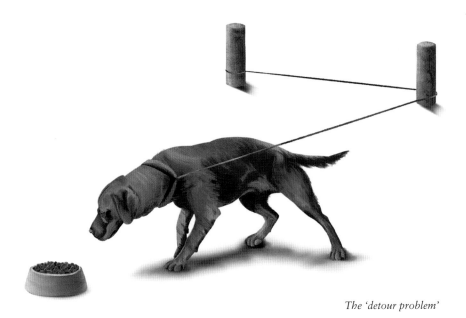

The 'detour problem'

But now, since you are human with a human power of imagination, you are probably simulating in your brain an unusually bright baboon. He sees himself, in his own imagination, pulling the plank carefully inwards, complete with bananas. Or reaching out with a stick and nudging the bananas along the plank towards him. Probably only highly evolved brains are capable of such simulations. Even dogs (above) perform surprisingly poorly on the so-called 'detour problem'. But if he succeeds, this imaginative baboon risks no pain and doesn't fall to his death but does it all by internal simulation. He simulates the fall in his imagination, and consequently refrains from venturing out along the plank. He then simulates the safe solution to the problem and gets the bananas.

I need hardly say that internal simulation of dangerous futures is preferable to the actual actions. Provided, of course, that the simulation leads to accurate prediction. Aircraft designers find it cheaper and safer to test model wings in wind tunnels rather than actual wings on real aeroplanes. And even wind tunnel models are more expensive than computer simulations or analytical calculations, if these can be done. Simulation still leaves some room for uncertainty.

The maiden flight of a new plane is still an informative event, however rigorously its parts have been subjected to ordeal by wind tunnel or computer simulation.

Once a sufficiently elaborate simulation apparatus is in place in a brain, emergent properties spring up. The brain that can imagine how alternative futures might affect survival can also, in the skull of a Dante or a Hieronymus Bosch, imagine the torments of Hell. The neurons of a Dalí or an Escher simulate disturbing images that will never be seen in reality. Non-existent characters come alive in the head of the great novelist and in those of her readers. Albert Einstein, in imagination, rode a sunbeam to his place among the immortals with Newton and Galileo. Philosophers imagine impossible experiments – the brain in a vat ('Where am I?'), atom-for-atom duplication of a human (which 'twin' would claim the 'personhood'?). Beethoven imagined, and wrote down, glories that he tragically could never hear. The poet Swinburne happened upon a forsaken garden on a sea cliff, and his imagination revived a pair of long-dead lovers whose eyes went seaward, 'a hundred sleeping years ago'. Keats reconstructed the 'wild surmise' with which stout Cortez and all his men stared at the Pacific, 'silent upon a peak in Darien'.

The ability to perform such feats of imagination sprang, emergently, from the Darwinian gift of vicarious internal simulation within the safe confines of the skull, of predicted alternative actions in the unsafe real world outside. The capacity to imagine, like the capacity to learn by trial and error, is ultimately steered by genes, by naturally selected DNA information, the genetic book of the dead.

8

The Immortal Gene

The central idea of *The Genetic Book of the Dead* grows out of a view of life that may be called the gene's-eye view. It has become the working assumption of most field zoologists studying animal behaviour and behavioural ecology in the wild, but it has not escaped criticism and misunderstanding, and I need to summarise it here because it is central to the book.

There are times when an argument can helpfully be expressed by contrast with its opposite. Disagreement that is clearly stated deserves a clear reply. I could hypothetically invent the opposite of the gene's-eye view, but fortunately I don't need to because the diametric opposite has been put, articulately and clearly, by my Oxford colleague (and incidentally my doctoral examiner, on a very different subject long ago) Professor Denis Noble. His vision of biology is alluring, and is shared by others whose expression of it is less explicit and less clear. Noble is clear. He ringingly hits a nail on the head, but it's the wrong nail. Here is his lucid and unequivocal statement, right at the beginning of his book *Dance to the Tune of Life*:

> This book will show you that there are no genes 'for' anything.
> Living organisms have functions which use genes to make

the molecules they need. Genes are used. They are not active causes.

That is precisely and diametrically wrong, and it will be my business in this chapter to show it.

If genes are not active causes in evolution, almost all scientists now working in the fields known as Behavioural Ecology, Ethology, Sociobiology, and Evolutionary Psychology have been barking up a forest of wrong trees for half a century. But no! 'Active causes' is precisely what genes must be: necessarily so if evolution by natural selection is to occur. And, far from being used by organisms, genes use organisms. They use them as temporary vehicles, which they exploit in the service of journeying to future generations. This is not a trivial disagreement, no mere word game. It is fundamental. It matters.

A physiologist of distinction, Denis Noble is captivated by the shattering complexity of the organism, of every last one of its trillions of cells. He sets out to impress his readers with the intricate co-dependency of all aspects of the living organism. As far as this reader is concerned, he succeeds. He sees every part as working inextricably with every other part in the service of the whole. In that service – and this is where he goes wrong – he sees the DNA in the nucleus of a cell as a useful library to be drawn upon when the cell needs to make a particular protein. Go into the nucleus, consult the DNA library there, take down the manual for making the useful protein, and press it into service. I devised that characterisation of Noble's position during a public debate with him in Hay-on-Wye, and he vigorously nodded his assent. DNA, in Noble's view, is the servant of the organism, in just the same way as the heart or the liver or any cell therein. DNA is useful to make a particular enzyme when you need it, just as the enzyme is useful for speeding up a chemical reaction … and so on.

Dance to the Tune of Life has the subtitle 'Biological Relativity'. Noble's usage of 'relativity' has only a tenuous and contrived connection with Einstein's, but it exactly matches that of the historian Charles Singer in *A Short History of Biology*:

The doctrine of the relativity of functions is as true for the gene as it is for any of the organs of the body. They exist and function only in relation to other organs.

Now here is Noble some ninety years later. He has the advantage over Singer in that we now know genes are DNA. But his sentiment about biological relativity, in conjunction with the quotation above, resonates perfectly with Singer's.

The principle of Biological Relativity is simply that there is no privileged level of causation in biology.

I shall argue that, no matter how complicatedly interdependent the parts of a living organism are when we are talking physiology, when we move to the special topic of evolution by Darwinian natural selection there is one privileged level of causation. It is the level of the gene. To justify that is the main purpose of this chapter.

Here's Singer's whole vitalistic passage from which I took the above quotation. It's the peroration of his book and is a perfect prefiguring of Noble's 'relativity'.

Further, despite interpretations to the contrary, the theory of the gene is not a 'mechanist' theory. The gene is no more comprehensible as a chemical or physical entity than is the cell or, for that matter, the organism itself. Further, though the theory speaks in terms of genes as the atomic theory speaks in terms of atoms, it must be remembered that there is a fundamental distinction between the two theories. Atoms exist independently, and their properties as such can be examined. They can even be isolated. Though we cannot see them, we can deal with them under various conditions and in various combinations. We can deal with them individually. Not so the gene. It exists only as a part of the chromosome, and the chromosome only as part of a cell. If I ask for a living chromosome, that is, for the only effective kind of chromosome, no one can give it to me except in its living surroundings any more than he can give me a living arm or leg. The doctrine of the relativity of functions is as true for the gene as it is for any of the organs of the body. They exist and function only in relation to other organs. Thus the last of the biological theories leaves us where the first started, in the presence of a power called life or psyche which is not only of its own kind but unique in each and all of its exhibitions.

Watson and Crick blew that out of the water in 1953. The triumphant field of digital genomics that they initiated falsifies every single one of Singer's sentences about the gene. It is true but trivial that a gene is impotent in the absence of its natural milieu of cellular chemistry. Here's Noble again, bringing Singer up to date but agreeing with his sentiment:

> There really is nothing alive in the DNA molecule alone. If I could completely isolate a whole genome, put it in a petri dish with as many nutrients as we may wish, I could keep it for 10,000 years and it would do absolutely nothing other than to slowly degrade.

Obviously a gene in a petri dish cannot do anything, and it would degrade as a physical molecule within months, let alone 10,000 years. But the *information* in DNA is potentially immortal, and causally potent. And that is the whole point. Never mind the physical molecule and never mind the petri dish. Let the sequence of A, T, C, G triplet codons of an organism's genome be written on a long paper scroll. Or, no, paper is too friable. To last 10,000 years, carve the letters deep in the hardest granite. To be sure, world-spanning ranges of highland massif would still be too small, but that is a superficial difficulty. In 10,000 years, if scientists still walk the Earth, they will read the sequence and type it into a DNA-synthesising machine such as we already have in early form. They'll have the embryological knowhow to create a clone of whoever donated the genome in the first place (just a version of the way Dolly the sheep was made). Of course, the DNA information would need the biochemical infrastructure of an egg cell in a womb, but that could be provided by any willing woman. The baby she bears, an identical twin of its 10,000-year dead predecessor, would be living repudiation of Singer and Noble.

That the information necessary to create the twin could be carved in lifeless granite and left for 10,000 years is a truth that fills me with amazement still, even seventy years after Watson and Crick prepared us for it. Charles Singer would be forced to recant his vitalism, while Charles Darwin, I suspect, would exult.

The point is that, transitory though *physical* DNA molecules themselves may be, the *information* enshrined in the nucleotide sequence is potentially eternal. Essential though the surrounding machinery is – messenger RNA, ribosomes, enzymes, uterus and all – they can be provided anew by any woman. But the information in an individual's DNA is unique, irreplaceable, and potentially immortal. Carving it in granite is a way to dramatise this. But it's not the practical way. In the normal course of events, DNA information achieves its immortality through being copied. And copied. And copied. Copied indefinitely, potentially eternally, down the generations. Of course, DNA can't copy itself on its own. Obviously, just as a computer disc can't copy itself without supporting hardware, DNA needs an elaborate infrastructure of cellular chemistry. But of all the molecules that are involved in the process, however essential they may be for the copying process, only DNA is actually copied. Nothing else in the body is so honoured. Only the information written in DNA.

You might think every part of the body is replicated. Does not every individual have arms and kidneys, and are these not renewed in every generation? Yes, but you'd be utterly wrong if you called it replication in the sense that genes are replicated. Arms and kidneys don't replicate to make new arms and kidneys. Here's the acid test, and it really matters. Make a *change* to an arm, say by a fracture or by pumping iron, and the change is *not* propagated to the next generation. Make a change in a germline gene, on the other hand, and the mutation may long outlast 10,000 years, copied again and again down the generations.

Before the invention of printing, biblical scriptures were painstakingly copied by scribes at regular intervals to forestall decay. The papyrus might crumble but the information lived on. Scrolls don't replicate themselves. They need scribes, and scribes are complicated, just as the enzymes involved in DNA replication are complicated. Through the mediation of scribes/enzymes information in scrolls/ DNA is copied with high fidelity. Actually, scribes might copy with lower fidelity than DNA replication can achieve. With the best will in the world human copyists make errors, and some zealous scribes were not above a little well-meant improvement. Older manuscripts of *Mark* 9, 29 quote Jesus as saying that a particular kind of demonic possession can be cured only by prayer. Later versions of the text, not content with mere prayer, say 'prayer and fasting'. It seems that some zealous scribe, perhaps belonging to a monkish order that especially valued fasting, thought to himself that Jesus must surely have *meant* to mention fasting, how could he not? So it was scarcely taking a liberty to put the words into his mouth. DNA is capable of higher fidelity of replication than that, but even DNA is not perfect. It does make mistakes – mutations. And in one important respect, DNA is unlike the over-zealous scribe: mutation is never biased towards improvement. Mutation has no way to judge in which direction improvement lies. Improvement is judged retrospectively. By natural selection.

So the information in DNA is potentially eternal even though the physical medium of DNA is finite. And let me repeat why this matters. Only the information contained in DNA is destined to outlive the body. Outlive in a very big way. Most animals die in a matter of years if not months or weeks. Few survive the ravages of decades, almost none centuries. And their physical DNA molecules die with them. But the information in the DNA can last indefinitely. I once attended an evolution conference in America where, at the farewell dinner, we were all challenged to produce an appropriate poem. My limerick ran as follows:

An itinerant Selfish Gene
Said 'Bodies a-plenty I've seen.
You think you're so clever
But I'll live for ever:
You're just a survival machine.'

And I raided Rudyard Kipling for the body's reply:

What is a body that first you take her,
Grow her up and then forsake her,
To go with the old Blind Watchmaker.

I have emphasised the immortality of the gene in the form of copies. But how big is the unit that enjoys such immortality? Not the whole chromosome: it is far from immortal. With minor exceptions such as the Y-chromosome, our chromosomes don't march intact down the centuries. They are sundered in every generation by the process of *crossing over.* For the purposes of this argument, the length of chromosome that should be considered significant in the long run depends upon how many generations it is allowed, by crossing over, to remain intact, when measured against the relevant selection pressures. I expressed this only slightly facetiously in my first book, *The Selfish Gene*, by saying that the title strictly should have been *The slightly selfish big bit of chromosome and the even more selfish little bit of chromosome.* A small fragment of chromosome, such as a gene responsible for programming one protein chain, can last 10,000 years. In the form of copies. But only fragments that are successful in negotiating the obstacle course that is natural selection actually do that. It's arguable that a better book title would have been *The Immortal Gene*, and I have adopted it as the title of this chapter. As we shall see in Chapter 12, it is no paradox that *The Cooperative Gene* would also have been appropriate.

How does a gene earn 'immortality'? In the form of copies, it influences a long succession of bodies so that they survive and reproduce, thereby handing the successful gene on to the next generation and potentially the distant future. Unsuccessful genes tend to disappear from the population, because the bodies they successively inhabit fail to survive into the next generation, fail to reproduce. Successful genes are those with a statistical tendency to inhabit bodies that are good at surviving and reproducing. And they enjoy that statistical tendency, positive or negative, by virtue of the *causal* influence they exert over bodies. So, we have arrived at the reason why it was profoundly wrong to say that genes are not active causes. Active causes is precisely and indispensably what they must be. If they were not, there could be no natural selection and no adaptive evolution.

'Cause' has a testable meaning. How do we ever identify a causal agent in practice? We do it by experimental intervention. Experimental intervention is necessary, because correlation does not imply causation. We remove, or otherwise manipulate, the putative cause, and we strictly must do so at random, a large number of times. Then we look to see whether there tends to be a statistically significant change in the putative effect. To take an absurd example, suppose we notice that the church clock in the village of Runton Acorn reliably chimes immediately after that of Runton Parva. If we're very naive, we jump to the conclusion that the earlier chiming causes the later. But of course it's not good enough to observe a correlation. The only way to demonstrate causation is to climb up the church tower in Runton Parva and manipulate the clock. Ideally, we force it to chime at random moments, and we repeat the experiment many times. If the correlation with the Runton Acorn chiming is maintained, we have demonstrated a causal link. The important point is that causation is demonstrated only if we *manipulate* the putative cause, repeatedly and at random. Of course, nobody would be silly enough to actually do this particular experiment with the church clocks. The result is too obvious. I use it only to clarify the meaning of 'cause'.

Now back to Denis Noble's statement that 'Genes are used. They are not active causes.' By our 'church clock' definition, genes most definitely are active causes because, if a gene mutates (a random change), we consistently observe a change in the body of the next generation – and subsequent generations for the indefinite future.

Mutation is equivalent to climbing the Runton Parva tower and changing the clock. By contrast, if there is a non-genetic change in the body (a scar, a lost leg, circumcision, an exaggeratedly muscular arm due to exercise, a suntan, acquired fluency in Esperanto or virtuosity on the bassoon), we do *not* observe the same thing in the next generation. There is no causal link.

Genetic information, then, is potentially immortal, is causal, and there's a telling difference between potentially immortal genes that succeed in being actually immortal and potentially immortal genes that fail. The reason some succeed and others fail is precisely that they have a *causal* influence, albeit a statistical one, on the survival and reproductive prospects of the many bodies that they inhabit, through successive generations and across many bodies through populations. It's important to stress 'statistical'. One copy of a good gene may fail to survive to the next generation because the body it inhabits is struck by lightning or otherwise suffers bad luck. More relevantly, one copy of a good gene may happen to find itself sharing a body with bad genes, and is dragged down with them. Statistics enter in because sexual recombination sees to it that good genes don't *consistently* share bodies with bad genes. If a gene is consistently found in bodies that are bad at surviving, we draw the statistical conclusion that it is a bad gene. After 10,000 years of recombining, shuffling, recombining again, a gene that remains in the gene pool is a gene that is good at building bodies: in collaboration with the other genes that it tends to share bodies with, and that means the other genes in the gene pool of the species (you may remember from Chapter 1 that the species can be seen as an averaging computer).

In *The Selfish Gene*, I used the image of the Oxford vs Cambridge Boat Race, the parable of the rowers. Eight oarsmen and a cox all have their part to play, and the success of the whole boat depends upon their cooperation. They must not only be strong rowers, they must be good cooperators, good at melding with the rest of the crew. The rowers, of course, represent genes, and they are arrayed along the length of the boat, as genes are arrayed along a chromosome. It's hard to separate the roles of the individual oarsmen, so intimate is their cooperation, and so vital is cooperative pulling together for the success of the whole boat. The coach swaps individual rowers in and out of his trial crews. Although it's hard to judge individual performance by watching them, he notices that certain individuals consistently seem to be members of the fastest trial crews. Other individuals consistently are seen to be members of slower crews. Although single individuals never row on their own, in the long run the best rowers show their mettle in the performance of the successive boats in which they sit.

Natural selection sorts out the good genes from the bad, precisely because of the *causal* influence of genes on bodies. The practical details vary from species to species. Genes that make for good swimmers are 'good genes' in a dolphin gene pool but not in a mole gene pool. Genes that make for good diggers are 'good genes' in a mole, wombat, or aardvark gene pool but not in a dolphin or salmon gene pool. Genes for expert climbing flourish in a monkey, squirrel, or chameleon gene pool but not in a swordfish, rhinoceros, or earthworm gene pool. Genes for aerodynamic proficiency flourish in a swallow or bat gene pool though not in a hippo or alligator gene pool.

But however varied the details of 'good' and 'bad' may be from species to species, the central point remains. Depending on their causal influence on bodies, genes either survive or don't survive to the next generation, and the next, and the next ... *ad infinitum*. Let me put it more forcefully: any Darwinian process, anywhere in the universe – and I'm pretty sure if there's life elsewhere in the universe it will be Darwinian life – any Darwinian process depends on trans-generational replicated information, and that information must have a causal influence on its probability of being replicated from one generation to the next. It happens that on our planet the replicated information, the causal agent in the Darwinian process, is DNA. It is wrong, utterly, blindingly, flat-footedly, downright wrong, to deny its fundamental role as a *cause* in the evolutionary process.

Have I laboured the point excessively? Would that it were excessive, but unfortunately there is reason to think that views such as those I have criticised here have been widely influential. Stephen Jay

Gould (whose errors were consistently masked by the graceful eloquence with which he expressed them) went so far as to reduce the role of genes in evolution to mere 'bookkeeping'. The metaphor of the bookkeeper has a dramatic appeal so seductive that it evidently seduced Gould himself. But it's as wide of the mark as it is possible to be. It is the bookkeeper's role to keep a passive record of transactions after they happen. When the bookkeeper makes an entry in his ledger, the entry does not cause a subsequent monetary transaction. It is the other way around.

I hope the preceding pages have convinced you that 'bookkeeping' is worse than a hollow travesty of the central causal role that genes play in evolution. It is the exact opposite of the truth, a metaphor as deeply wrong as it is superficially persuasive. Gould was also a proponent of 'multi-level selection', and this is another respect in which he is seen as an opponent of the gene's-eye view of evolution (see, for instance, the philosopher Kim Sterelny's perceptive book *Dawkins Versus Gould: Survival of the Fittest*). Gould, and others, insisted that natural selection occurs at many levels in the hierarchy of life: species, group, individual, gene. The first thing to say about this is that although there is a persuasive hierarchy, a real ladder, the gene doesn't belong on it. Far from being the bottom rung of a ladder, far from being on the ladder at all, the gene is set off to one side. Precisely because of its privileged role as a causal agent in evolution. The gene is a replicator. All other rungs in the ladder are *vehicles*, a term that I shall explain later in this chapter.

As for higher levels of selection, there is, to be sure, a sense in which some species survive at the expense of others. This can look a bit like natural selection at the species level. The native red squirrel in Britain is steadily going extinct as a direct result of the lamentable whim of the 11th Duke of Bedford in the nineteenth century to introduce American grey squirrels. The greys out-compete the smaller reds, and also infect them with squirrel pox, to which they themselves have evolved resistance over many generations in America. Ecological replacement of a species by a competitor species looks superficially like natural selection. But the resemblance is empty and misleading.

This kind of 'selection' does not foster evolutionary adaptation. It's not natural selection in the Darwinian sense. You would not say that any aspect of the grey squirrel's body or behaviour was a device to drive red squirrels extinct, whereas you might happily talk about the Darwinian function of its bushy tail, meaning those aspects of the tail that assisted ancestral squirrels to out-compete rival squirrel individuals of the same species, with a slightly different tail.

In 1988, I published a paper called 'The Evolution of Evolvability'. This is the closest I have come to supporting something like 'multi-level selection'. My thesis was that certain body plans, for example the segmented body plans of arthropods, annelids, and vertebrates, are more 'evolvable' than others. I quote from that paper:

> I suspect that the first segmented animal was not a dramatically successful individual. It was a freak, with a double (or multiple) body where its parents had a single body. Its parents' single body plan was at least fairly well-adapted to the species' way of life, otherwise they would not have been parents. It is not, on the face of it, likely that a double body would have been better adapted … What is important about the first segmented animal is that its descendant lineages were champion *evolvers*. They radiated, speciated, gave rise to whole new phyla. Whether or not segmentation was a beneficial adaptation during the individual lifetime of the first segmented animal, segmentation represented a change in embryology that was pregnant with evolutionary potential.

I envisioned that my concept of 'evolvability' should be regarded as a property of embryology. Thus, a segmented embryology has high evolvability potential, meaning an embryology that lends itself to rich evolutionary divergence. The world tends to become populated by clades with high evolvability potential. A clade is a branch of the tree of life, meaning a group plus its shared ancestor. 'Birds' constitutes a clade, for all birds have a single common ancestor not shared by any non-birds. 'Fish' is not a clade, because the common ancestor of all fish is shared by all terrestrial vertebrates including us, who are not fish. 'Mammals' is a clade, but only if you include so-called 'mammal-like reptiles'. It would be unhelpful and confusing to call the evolution of evolvability group selection. 'Clade selection', a coining of George C Williams, fits the bill.

What other criticisms of the gene's-eye view should we consider? Many would-be critics have pointed out that there is no simple one-to-one mapping between a gene and a 'bit' of body. Though true, that's not a valid criticism at all, but I need to explain it because some people think it is. You know those gruesome butchers' diagrams, where a map of a cow's body is defaced by lines representing named 'cuts' of meat: 'rump', 'brisket', 'sirloin', etc? Well, you can't draw a map like that for domains of genes. There's no 'border' you can draw on the body, marking where the 'territory' of one gene ends and that of the next one begins. Genes don't map onto bits of body; they map onto timed embryological *processes*. Genes influence embryonic development, and a *change* in a gene (mutation) maps onto a *change* in a body. When geneticists notice a gene's effects, all they are really seeing is a *difference* between individuals that have one version ('allele') of the gene and individuals that don't. The units of phenotype that geneticists count, or trace through pedigrees, traits such as the Hapsburg jaw, albinism, haemophilia, or the ability to smell freesias, loop the tongue, or disperse the froth on contact with beer, are all identified as *differences* between individuals. For, of course, countless genes are involved in the development of any jaw, Hapsburg or not; any tongue, loopy or not. The Hapsburg jaw gene is no more than a gene for a *difference* between some individuals and other individuals. Such is the true meaning whenever anyone talks of a gene 'for' anything. Genes are 'for' individual differences. And, just as the eyes of a geneticist are focused on individual differences in phenotype, so also, precisely and acutely, are the eyes of natural selection: differences between those who have what it takes to survive and those who don't.

As for the all-important *interactions* between genes in influencing phenotype, here's a better metaphor than the butcher's map. A large

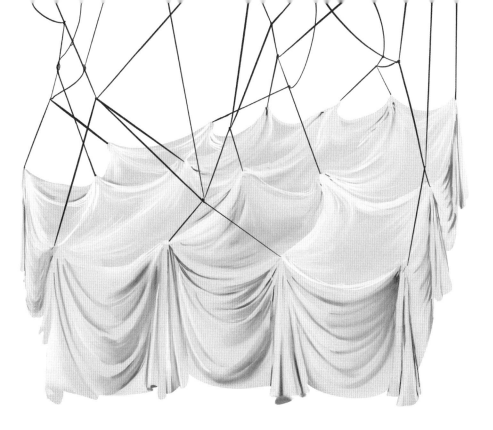

A balance of tensions

sheet hangs from the ceiling, suspended from hooks by hundreds of strings attached to different places all over the sheet. It may help the analogy to consider the strings as elastic. The strings don't hang vertically and independently. Instead, they can run diagonally or in any direction, and they interfere with other strings by cross-links rather than necessarily going straight to the sheet itself. The sheet takes on a bumpy shape, because of the interacting tensions in the tangled cat's-cradle of hundreds of strings. As you've guessed, the shape of the sheet represents the phenotype, the body of the animal. The genes are represented by the tensions in the strings at the hooks in the ceiling. A mutation is either a tug towards the hook or a release, perhaps even a severing of the string at the hook. And, of course, the point of the parable is that a mutation at any one hook affects the *whole balance of tensions* across the tangle of strings. Alter the tension at any one hook, and the shape of the whole sheet shifts. In keeping with the sheet model, many, if not most, genes have 'pleiotropic' (multiple) effects, as defined in Chapter 4.

For practical reasons, geneticists like to study the minority of genes that do have definable, seemingly singular effects, like Gregor Mendel's smooth or wrinkled peas, for example. But even such 'major genes' often have a surprisingly miscellaneous collection of other pleiotropic effects, sprinkled seemingly at haphazard around the body. And it's not surprising that this should be so: genes exert their effects at many stages of embryonic development. It's only to be expected, therefore, that they'll have pleiotropic consequences even at opposite ends of the body. A change in tension at one hook leads to a comprehensive shapeshift, all over the whole sheet.

There's no one-to-one mapping, then, from single gene to single 'bit' of body. We have no butcher's map here. But not by a jot or even a tittle does this fact threaten the gene's-eye view of evolution. However pleiotropic, however complicated and interactive the effects of a gene may be, you can still add them all up to derive a net positive or net negative effect of a *change* (mutation) in its influence on the body: a net effect on its chances of surviving into the next generation. Such causal influences on a gene's own survival in the gene pool come unscathed through the complications, notwithstanding numerous interactions with other genes – the other genes with which it jointly affects the tensions in all the strings. When the gene in question mutates, the whole shape of the sheet may shift, with perhaps lots of pleiotropic changes all over the body. But the net effect of all these changes, in different parts of the body, and in interaction with many other genes, must be either positive or negative (or neutral) with respect to survival and reproduction. That is natural selection.

The tension in the genetic strings is affected too by environmental influences. See these as yet more strings tugging from the side, rather than from hooks in the ceiling. The developing animal is, of course, influenced by the environment as well as by the genes, always in interaction with the genes. But again, this doesn't matter one iota to the gene's-eye view of evolution. To the extent that, under available environmental conditions, a *change* in a gene causes a *change* in that gene's chances of making it through the generations (either positive or negative), natural selection will occur. And natural selection is what the gene's-eye view is all about.

So much for that criticism of the gene's-eye view. What else do we have? Granted that genes are active causes in evolution, it is the whole individual body that we observe behaving as an active agent. This fact, too, is often wrongly seen as a weakness of the gene's-eye view. Yes, of course, it is the whole animal who possesses executive instruments with which to interact with the world – legs, hands, sense organs. It's the whole animal who restlessly searches for food, trying first this avenue of hope, then switching to another, showing all the symptoms of questing appetite until consummation is reached. It is the individual animal who shows fear of predators, looks vigilantly up and around, jumps when startled, runs in evident terror when pursued. It is the individual animal who behaves as a unitary agent when courting the opposite sex. It is the individual animal who skilfully builds a nest, and works herself almost to death caring for her young.

The animal, the individual animal, the whole animal, is indeed an agent, striving towards a purpose, or set of purposes. Sometimes the purpose seems to be individual survival. Often it is reproduction and the survival of the individual's children. Sometimes, especially in the social insects, it is the survival and reproduction of relatives other than children – sisters and nieces, nephews and brothers. My late colleague WD Hamilton (he of the palimpsest postcard in Chapter 1) formulated the general definition of the exact mathematical quantity that an individual under natural selection is expected to maximise as it engages in its purposeful striving. It includes individual survival. It includes reproduction. But it includes more, because genes are shared with collateral relatives, and gene survival can therefore be fostered by enabling the survival and reproduction of a sister or a nephew. He gave a name to the exact quantity that an individual organism should strive to maximise: 'inclusive fitness'. He condensed his difficult mathematics into a long and rather complicated verbal definition:

Inclusive fitness may be imagined as the personal fitness which an individual actually expresses in its production of adult offspring as it becomes after it has been first stripped and then augmented in certain ways. It is stripped of all components which can be considered as due to the individual's social environment, leaving the fitness which he would express if not exposed to any of the harms or benefits of that environment. This quantity is then augmented by certain fractions of the quantities of harm and benefit which the individual himself causes to the fitnesses of his neighbours. The fractions in question are simply the coefficients of relationship appropriate to the neighbours whom he affects: unity for clonal individuals, one-half for sibs, one-quarter for half sibs, one-eighth for cousins ... and finally zero for all neighbours whose relationship can be considered negligibly small.

Pretty convoluted? A bit hard to read? Well, it has to be convoluted because inclusive fitness is a hard idea. It's necessarily convoluted in my view because looking at it from the individual's point of view is an unnecessarily convoluted way of thinking about Darwinism. It all becomes blessedly simple if you dispense with the individual organism altogether and go straight to the level of the gene. Bill Hamilton himself did this in practice. In one of his papers, he wrote:

> let us try to make the argument more vivid by attributing to the genes, temporarily, intelligence and a certain freedom of choice. Imagine that a gene is considering the problem of increasing the numbers of its replicas, and imagine that it can choose between causing purely self-interested behaviour by its bearer ... and causing 'disinterested' behaviour that benefits in some way a relative.

See how clear and easy to follow that is, compared to the previous quotation on inclusive fitness. The difference is that the clear passage adopts the gene's-eye view of natural selection. The difficult passage is what you get when you re-express the same idea from the point of view of the individual organism. Hamilton gave his blessing to my half-humorous informal definition: 'Inclusive fitness is that quantity that an individual will appear to be maximising, when what is really being maximised is gene survival.'

	Role	Maximises
Gene	Replicator	Survival
Organism	Vehicle	Inclusive fitness

Bill Hamilton

The individual organism, in my terminology, is a 'vehicle' for survival of copies of the 'replicators' that ride inside it. The philosopher David Hull got the point after an extensive correspondence with my then student Mark Ridley, but he substituted the word 'interactor' for my 'vehicle'. I never quite understood why. Depending on your preference you can see either the vehicle or the replicator as the agent that maximises some quantity. If it's the vehicle, then the quantity maximised is inclusive fitness, and rather complicated. But equivalently, if it's the replicator, the quantity maximised is simple: survival. I don't want to downplay the importance of vehicles as units of action. It is the individual organism who possesses a brain to take decisions, based on information supplied by senses, and executed by muscles. The organism ('vehicle') is the unit of action. But the gene ('replicator') is the unit that survives. On the gene's-eye view, the very existence of vehicles should not be taken for granted but needs explaining in its own right. I essayed a kind of explanation in 'Rediscovering the Organism', the final chapter of *The Extended Phenotype*.

Replicators (on our planet, stretches of DNA) and vehicles (on our planet, individual bodies) are equally important entities, equally important but they play different, complementary roles. Replicators may once have floated free in the sea but, to quote *The Selfish Gene*, 'they gave up that cavalier freedom long ago. Now they swarm in huge colonies, safe inside gigantic lumbering robots' (individual bodies, vehicles). The gene's-eye view of evolution does not play down the role of the individual body. It just insists that that role ('vehicle') is a different kind of role from that of the gene ('replicator').

Successful genes, then, survive in bodies down the generations, and they *cause* (in a statistical sense) their own survival by their 'phenotypic' effects on the bodies that they inhabit. But I went on to amplify the gene's-eye view by introducing the notion of the *extended* phenotype. For the causal arrow doesn't stop at the body wall. Any causal effect on the world at large – any causal effect that can be attributed to the presence of a gene as opposed to its absence, and that influences the gene's chances of survival, may be regarded as a phenotypic effect, of Darwinian significance. It has only to exert some kind of statistical influence on the chances, positive or negative, on that gene's surviving in the gene pool. I must now revisit the extended phenotype, for it is, to me, an important part of the gene's-eye view of evolution.

 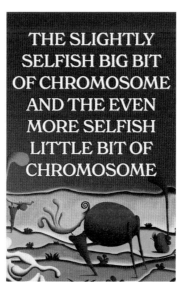

Alternative titles for The Selfish Gene, *all true to its content*

9

Out Beyond the Body Wall

Imagine the furore if Jane Goodall reported seeing chimpanzees building an amazing stone tower in a forest clearing. They meticulously select stones of the correct shape for the purpose, rotating each one until it snugly fits neighbouring stones. Then the chimps cement it securely in place before picking out another stone. They evidently like to use two radically different sizes of stones, small ones to build the walls themselves, and much larger ones to provide outer fortification and structural strength, the all-important supporting walls. The discovery would be a sensation, headline news, the subject of breathless BBC discussions. Philosophers would jump on it, there'd be passionate debates about personhood, moral rights, and other topics of philosophical moment. The tower is ill-suited to housing its builders. If not functional, then, is it some kind of monument? Does it have ritual or ceremonial significance like Stonehenge? Does the tower show that religion is older than mankind? Does it threaten the uniqueness of man?

The edifice pictured is a real animal construction, but not one built by chimpanzees; the reality is much smaller, and it doesn't stand up like a monument but lies flat on the bottom of a stream. It is the house of a little insect, the larva of a caddis fly, *Silo pallipes*. Caddis adults fly in search of mates and live only a few weeks, but their larvae grow for up to two years under water, living in mobile homes that they build for themselves out of materials gathered from their surroundings, cementing them with silk that they secrete from glands in the head. In the case of *Silo pallipes* (see top left of picture on page 199) the building material is local stone. Its astonishing building skills were unravelled by Michael Hansell, now our leading expert on animal architecture in general.

These larvae are master masons. Just look at the delicate placing of the small stones between the carefully chosen large ones buttressing the sides. Hansell showed how they select stones, choosing by size and shape but not by weight. Ingenious experiments in which

If only chimps had the skills of a caddis larva...

he removed parts of the house showed how the larvae fit appropriate stones in the gaps, and cement them in place. Just as impressive is the log house at top right of the picture. This was built not by a caddis larva but by a caterpillar, a so-called bagworm. Caddises in water and bagworms on land have converged independently on the habit of building houses from materials that they gather from their surroundings. The picture shows a selection of caddis and bagworm houses.

The word 'phenotype' is used for the bodily manifestation of genes. The legs and antennae, eyes and intestines are all parts of the caddis's phenotype. The gene's-eye view of evolution regards the phenotypic expression of a gene as a tool by which the gene levers itself into the next generation – and, by implication, an indefinite number of future generations. What this chapter adds is the notion of the *extended* phenotype. Just as the shell of a snail is part of its phenotype, its shape, size, thickness, etc. being affected by snail genes, so the shape, size, etc. of a stone caddis house or twiggy bagworm cocoon are all manifestations of genes. Because these phenotypes are not part of the animal's own body, I refer to them as *extended* phenotypes.

These elegant constructions must be the products of Darwinian evolution, no less than the armoured body wall of a lobster, a tortoise, or an armadillo. And no less than your nose or big toe. This means they have been put together by the natural selection of genes. Such is the Darwinian justification for speaking of extended phenotypes. There must be genes 'for' the various details of caddis and bagworm houses. This means only that there must be, or have been, genes in the insects' cells, variants of which cause variation in the shape or nature of houses. To conclude this, we need assume only that these houses evolved by Darwinian natural selection, an assumption that no serious biologist would dispute, given their elegant fitness for purpose. The same is true of the nests of potter wasps, mud dauber wasps, and ovenbirds. Built of mud rather than living cells, they are extended phenotypes of genes in the bodies of the builders.

While their grasshopper cousins sing with serrated legs, male crickets sing with their wings, scraping the top of one front wing against a rough 'file' on the underside of the other front wing. Among

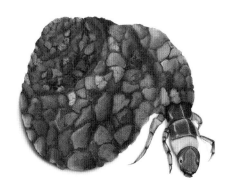

Potter wasp

Mud dauber

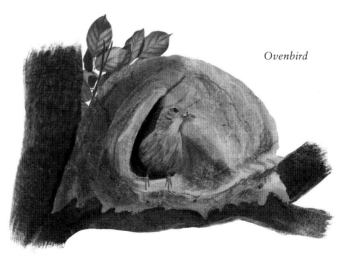

Ovenbird

their songs, the 'calling song' is loud enough to attract females within a certain radius, and to deter rival males. But what if it could be amplified, widening the catchment area for pulling females? Some kind of megaphone, perhaps? We use a megaphone as a simple directional amplifier, which works by 'impedance matching'. No need to go into what that means, except to say that, unlike an electronic amplifier, it adds no extra energy. Instead, it concentrates the available energy in a particular direction. Could a cricket grow a megaphone out of its horny cuticle – a phenotype in the conventional sense? Like the remarkable backwards-facing trombone of the dinosaur *Parasaurolophus*, which probably served as a resonator for its bellowings. Crickets could have evolved something like that. But an easier material was to hand, and mole crickets exploited it.

Parasaurolophus

Mole crickets, as their name suggests, are digging specialists. Their front legs are modified to form stout spades, strongly resembling those of moles, albeit on a smaller scale. The similarity, of course, is convergent. Some species of mole crickets are so deeply committed to underground life that they cannot fly at all. Given that a mole cricket could benefit from a megaphone, and given that it digs a burrow, what more natural than to shape the burrow as a megaphone? In the case of *Gryllotalpa vineae* it is a double megaphone, like an old-fashioned clockwork gramophone with two horns. Henry Bennet-Clark showed that the double horn concentrates the sound into a disc section rather than letting it dissipate in all directions as a hemisphere. Bennet-Clark was able to hear a single *Gryllotalpa vineae* (a species he discovered himself) from 600 metres away. The range of no ordinary cricket comes close.

Assuming it's as beautifully functional as it seems to be, the mole cricket's megaphone must have evolved by natural selection, as a step-by-step improvement, in just the same way as the digging hand

Mole cricket   Mole

or as any part of the cricket's own body. Therefore, there must be genes controlling horn shape, just as there are genes controlling wing shape or antenna shape. And just as there are genes controlling the patterning of cricket song itself. If there were no genes for horn shape, there would be nothing for natural selection to choose. Once again, remember that a gene 'for' anything is only ever a gene whose alternative alleles encode a difference between individuals.

Mole cricket with double megaphone burrow

Now, when contemplating the double megaphone (or, for that matter, the houses of caddises and bagworms) you might be tempted to say something along the following lines. Cricket burrows are not like wings or antennae. They are the product of cricket *behaviour*, whereas wings and antennae are anatomical structures. We are accustomed to the idea of anatomical structures being under the control of genes. Can the same be said of behaviour, of cricket digging behaviour, or the sophisticated stonemasonry behaviour of a caddis larva? Yes, of course it can. And there is nothing to stop it being said of artifacts that are produced by the behaviour. The artifacts are just one further step in the causal chain from gene to protein to … a long cascade of processes in the embryo, culminating in the adult body.

There are numerous studies of the genetics of behaviour, including, as it happens, the genetics of cricket song. I want to discuss this work because, weirdly, behaviour genetics arouses a scepticism never suffered by anatomical genetics. Cricket song (though not specifically mole cricket song) has been the subject of penetrating genetic research by David Bentley, Ronald Hoy, and their colleagues in America. They studied two species of field cricket, *Teleogryllus commodus* from Australia and *Teleogryllus oceanicus*, also Australian but found in Pacific islands too. Adult crickets who have been brought up in isolation from other crickets sing normally. Nymphs who have not yet undergone their final moult to adulthood never sing, but in the laboratory their thoracic ganglia can be induced to emit nerve impulses with a time-pattern identical to the species song pattern. These facts strongly suggest that the instructions for how to sing the species song are coded in the genes. And those genes must be relevantly different in the two species, for their song patterns are different. This is beautifully confirmed by hybridisation experiments.

In nature these two *Teleogryllus* species don't interbreed, but they can be induced to do so in the laboratory. The diagram, from Bentley and Hoy, shows the songs of the two species and of various hybrids between them. All cricket songs are made up of pulses separated

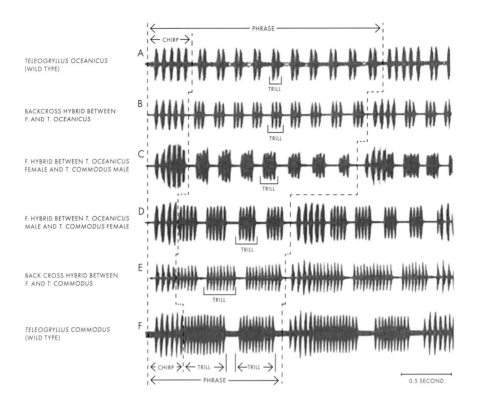

PHRASE

← CHIRP →

TELEOGRYLLUS OCEANICUS
(WILD TYPE)

A

TRILL

BACKCROSS HYBRID BETWEEN
F. AND *T. OCEANICUS*

B

TRILL

F. HYBRID BETWEEN *T. OCEANICUS*
FEMALE AND *T. COMMODUS* MALE

C

TRILL

F. HYBRID BETWEEN *T. OCEANICUS*
MALE AND *T. COMMODUS* FEMALE

D

TRILL

BACK CROSS HYBRID BETWEEN
F. AND *T. COMMODUS*

E

TRILL

TELEOGRYLLUS COMMODUS
(WILD TYPE)

F

← CHIRP → ← TRILL → ← TRILL →
← PHRASE →

0.5 SECOND

Songs of pure bred and hybrid crickets

by pauses. *T.oceanicus* (A in the picture) has a 'chirp' consisting of about five pulses followed by a series of about ten 'trills', each trill always made up of two pulses, closer to each other than the pulses of the chirp. We hear a rhythmic repetition pattern of trills. To my ears the trills sound slightly quieter than the chirps. After about ten of these double-pulse trills there's another chirp. And the cycle repeats rhythmically, over and over again indefinitely. *T.commodus* (F) has a similar pattern of alternating chirps and trills. But instead of a series of ten or so double-pulse trills, there is only one long trill or perhaps two, between chirps.

Now to the interesting question: what about the hybrids? Hybrid songs (C and D) are intermediate between those of the two parent species (A and F). It makes a difference which species is the male (compare C with D), but we needn't go into that here, interesting though it is for what it might tell us about sex chromosomes. In any case, hybrid song is a beautiful confirmation of genetic control of a behaviour pattern. Further evidence (B and E) comes from crossing hybrids with each of the two wild species (what geneticists call a backcross). If you compare all five songs, you'll note a satisfying generalisation: hybrid songs resemble the two wild species' songs in proportion to the number of genes the hybrid individual has inherited from each species. The more *oceanicus* genes an individual has, the more its song resembles wild *oceanicus* rather than *commodus*. And vice versa. As your eyes move down the page from *oceanicus* towards *commodus*, the more you detect resemblance to *commodus* song. This suggests that several genes of small effect ('polygenes') sum their effects. And what is not in doubt is that the species-specific song patterns that distinguish these two species of crickets are coded in the genes: a nice example of how behaviour is just as subject to genetic control as anatomical structures are. Why on earth shouldn't it be? The logic of gene causation is identical for both. Both are products of a chain of causation, with the behaviour having one more link in the chain.

You could do a similar study of the genetics of megaphone-building behaviour. But you might as well go to the next step in the causal chain, the megaphone itself. Do a genetic study of differences between megaphones. They are extended phenotypes of mole cricket genes. This has not been done, but nothing prevents it. Again, nobody has studied the genetics of caddis houses, but there's no reason why they shouldn't, although there might be practical difficulties in breeding them in the lab. Michael Hansell was once giving a talk at Oxford, on the building behaviour of caddis larvae. In passing, he was lamenting his failed attempts to breed caddises in the lab, for he wished he could study their genetics. At this, the Professor of Entomology growled from the front row: 'Haven't you trrrried cutting their heads off?' It seems that the insect brain exercises inhibitory influences such that beheading can be expected to have a releasing effect.

If you were to succeed in breeding caddises in captivity, you could systematically select changes in caddis houses over generations. Or you could artificially select for mole cricket megaphone size or shape, generation by generation, breeding from those individuals whose horns happen to be wider, or deeper, or of a different shape. You could breed giant megaphones, just as you might breed giant antennae or mandibles.

That would be artificial selection, but something like it must have happened through natural selection. Whether by artificial or natural selection, the evolution of larger megaphones could come about only by differential survival of genes for megaphone size. For the megaphone to have evolved in the first place as a Darwinian adaptation, there had to be genes for megaphone shape. The notion of the extended phenotype is a necessary part of the gene's-eye view of evolution. The extended phenotype should be an uncontroversial addition to Darwinian theory.

But aren't those 'genes for megaphone shape' *really* genes for altered digging behaviour, which is part of the 'ordinary' phenotype of the cricket? Aren't genes for caddis house shape 'really' genes for building behaviour, that is to say, 'ordinary' phenotypic manifestations within the body? Why talk about 'extended' phenotypes outside the body at all? Well, you could equally well say that the genes for altered digging behaviour are 'really' genes for changed wiring in the ganglia in the thorax. And genes for changes in the thoracic ganglia are, in turn, 'really' genes for changes in cell-to-cell interactions in embryonic development. And they, in turn, are 'really' ... and so on back until we hit the ultimate 'really'. Genes are really really *really* only genes for changed proteins, assembled according to the rules for translating the sixty-four possible DNA triplet codons into twenty amino acids plus a punctuation mark. I repeat, because it is important, we have here a chain of causation whose first steps (DNA codons choosing amino acids) are knowable, whose final step (megaphone shape) is observable and measurable, and whose intermediate steps are buried in the details of embryology and nerve connections – perhaps inscrutable but necessarily there. The point is that any one of those many intermediate steps in the chain of causation could be regarded as 'phenotype', and could be the target of selection, artificial or natural. There is no logical reason to stop the chain at the animal's body wall. Megaphone is 'phenotype', every bit as much as nerve-wiring is phenotype. Every one of those steps, both in the cricket's body and extended outside it, can be regarded as *caused* by gene differences. Just the same is true of the chain of causation leading from genes to caddis house, even though the behavioural step, the actual building itself, involves sophisticated trial and error in the selection of suitable stones and rotating them into position to fit the existing structure. And now to advance the argument a stage further. The extended phenotype of a gene can reach into the body of a different individual.

Natural selection doesn't see genes for digging behaviour directly, nor does it see neuron circuitry directly, nor indeed megaphone shape directly. It sees, or rather hears, song loudness. Gene selection is what ultimately matters, but song loudness is the proxy by which gene selection is mediated, via a long series of intermediates. But even song loudness is not the end of the causal chain. As far as natural selection is concerned, song loudness only matters insofar as it attracts females (and deters males, but let's not complicate the argument). The causal chain extends to a radius where it exerts an influence on a female cricket. This has to mean that a change in female behaviour is part of the extended phenotype of genes in a male cricket. Therefore, the extended phenotype of a gene can reside in another individual. The general point I am aiming towards is that the phenotypic expression of a gene can extend even to living bodies other than the body in which the genes sit. Just as we can talk of a gene 'for' a Hapsburg lip, or a gene 'for' blue eyes, so it is entirely proper to talk of a gene (in a male cricket) 'for' a change in another individual's behaviour (in this case a female cricket).

We saw in Chapter 7 that song in male canaries and ring doves has a dramatic effect on female ovaries. They swell hugely, with a corresponding rush of hormones and all that it entails. The consequent changes in female behaviour and physiology are in truth phenotypic expression of male genes. Extended phenotypic expression. You may deny it only if you deny Darwinian selection itself.

Ears are not the only portals into a female dove's brain through which a male's genes might exert an extended phenotypic influence. Male birds of many species glow with conspicuous colours. These cannot be good for individual survival, but they are still good for the survival of the genes that fashioned them. They achieve this good by assisting individual reproduction at the expense of individual survival. With few exceptions, it is males that sacrifice their personal longevity on the altar of gene survival, through sexually attractive coloration. In those species such as pheasants or birds of paradise, where males dazzle, females are usually drabber in colour, often well camouflaged. Bright coloration in males is favoured, either through attracting females or through besting rival males. In both cases, the naturally selected genes for bright coloration have extended phenotypic expression in the changed behaviour of other individuals. I don't know whether exposure to a male peacock fan causes peahen ovaries to change, as male dove bow-cooing song does to female dove ovaries. It wouldn't surprise me. I'd even be surprised if it didn't.

Unfortunately, predators tend to have eyes like the eyes of the females whom the male is seeking to impress. What is conspicuous to one will probably be conspicuous to all. It's worth it to the male, or rather to the genes that coloured him. Even if his finery costs him his life, it can already have paid its way in previous success with females. But is there some way a male bird could manipulate females via their eyes without calling attention to himself? Could he shed his dangerously conspicuous personal phenotype, offloading it to an extended phenotype at a safe distance from his own body? 'Shed' and 'offload', of course, must be understood over evolutionary time. We aren't talking about shedding feathers in an annual moult, although that happens too – perhaps for the same reason. Black-headed gulls, for instance, shed their conspicuously contrasting face masks as soon as the breeding season is over.

Bower birds are a family of birds inhabiting the forests of New Guinea and Australia. Their name comes from a remarkable and

unique habit. They build 'bowers' to seduce females. The skills needed
to build a bower could be seen as a distant derivative of nest-building
skills, and perhaps ultimately derived from them. But the bower is
emphatically not a nest. No eggs are laid in it, no chicks reared there.
Female bower birds build nests to house eggs as other birds do, and
their nests don't resemble male bowers.

The bower's sole purpose is to attract females, and males take
enormous pains in their creation. First, they clear stray leaves and
other debris from the arena in which the bower is to be built. Then
the bower itself is assembled from twigs and grass. The details vary

from species to species. Some resemble a Robinson Crusoe hat, some a grand archway, others a tower. The final stage of bower design is, I think, the most remarkable of all. The ground in front of and under the bower is colourfully and – I can't resist saying – tastefully decorated. The male gathers decorative objects – coloured berries, flowers, even bottle tops. Movies of male bower birds at work irresistibly remind me of an artist putting the finishing touches to a canvas, standing back, head cocked judgmentally, then darting forward to make a delicate adjustment, standing back again and surveying the effect with head on one side before darting forward again. That is what emboldened me to use a word like 'tastefully'. It is hard to resist the impression that the bird is exercising his aesthetic judgement in perfecting a work of art. Even if the decorated bower is not to every human's taste, or even every female bower bird's, the 'touching up' behaviour of the male almost forces the conclusion that the male has taste of his own, and he is adjusting his bower to meet it.

Remember the discussion in Chapter 7, where I suggested that when male songbirds learn to sing, they are exercising their own aesthetic judgement? The evidence shows, you'll remember, that young birds burble at random, choosing, by reference to a template, which random fragments to incorporate into their mature song. The male, I argued, has a similar brain to a female of his own species. Not surprisingly, therefore, whatever appeals to him can be expected to appeal to her. The development of song in the young bird could be regarded as a work of creative composition in which the male adopts the principle of 'whatever turns me on will probably appeal to a female too'. I see no reason to refrain from a similar aesthetic interpretation of bower-building. 'I like the look of a heap of blue berries just there. So there's a good chance that a female of my own species will like it too … And perhaps a single red flower over there … or, no, it looks better here … and better still, slightly to the left, and why not set it off with some red berries?' Of course, I am not literally suggesting that he thinks it through in so many words.

Species differ as to their preferred decoration colours, as well as the shape of their bowers. The satin bower bird (page 209) goes for blue, a fact that may be connected with the blue-black sheen of his plumage, or the species' brilliant blue eyes. The male satin bower bird who built this bower has discovered blue drinking straws and bottle tops, and laid out a rich feast of blue to delight the female eye. More soberly, the Great Bower Bird, *Chlamydera nuchalis*, says it with shells and pebbles (opposite).

The bower is an extended phenotype of genes in the body of the male bower bird. An external phenotype, which presumably has the advantage that its extravagance is not worn on the body and therefore will not call predators' attention to the male himself. I do not know whether exposure to a more than usually magnificent bower stimulates a hormone surge in the blood of a female, but again the research on ring doves and canaries would lead me to expect this.

We are accustomed to thinking of genes as being physically close to their phenotypic targets. Extended phenotypes can be large, and far distant from the genes that cause them. The lake flooded by a beaver's dam is an extended phenotype of beaver genes, extended in some cases over acres. The songs of gibbons can be heard a kilometre away in the forest, howler monkeys as much as five kilometres: true genetic 'action at a distance'. These vocalisations have been favoured by natural selection because of their extended phenotypic effect on other individuals. Chemical signals can achieve a great range among moths. Visual signals require an uninterrupted line of sight, but the principle of genetic action at a distance remains. The gene's-eye view of evolution necessarily incorporates the idea of the extended phenotype. Natural selection favours genes for their phenotypic effects, whether or not those phenotypic effects are confined to the body of the individual whose cells contain the genes.

In 2002, Kim Sterelny, editor of the journal *Biology and Philosophy*, marked the twentieth anniversary of the publication of *The Extended Phenotype* by commissioning three critical appraisals, plus a reply from me. The special issue of the journal came out in 2004. The criticisms were thoughtful and interesting, and I tried to follow suit in my reply, but all this would take us too far afield here. I concluded my piece with a humorously grandiose fantasy about the building of a future Extended Phenotypics Institute. This pipedream edifice was to have three wings, the Zoological Artifacts Museum (ZAM), the laboratory of Parasite Extended Genetics (PEG), and the Centre for Action at a Distance (CAD). The subjects covered by ZAM and CAD have dominated this chapter. PEG must wait till the final chapter. Parasites often exert dramatic extended phenotypic effects on their hosts, manipulating the host's behaviour to the parasite's advantage, often in bizarrely macabre ways. The parasite doesn't have to reside in the body of the host, so there is an overlap with CAD, the Action at a Distance wing. Cuckoo chicks are external parasites who exert extended phenotypic influence over the behaviour of their foster parents. And cuckoos are so fascinating they deserve a chapter of their own. For a different reason, now to be explained.

10

The Backward Gene's-Eye View

The previous two chapters constituted my short reprise of the gene's-eye view of evolution as I explained it in *The Selfish Gene* and *The Extended Phenotype*. I want, now and in the next chapter, to offer the gene's-eye view in another way, a way that is particularly suitable for *The Genetic Book of the Dead*. This is to imagine the view seen by a gene as it 'looks' *backwards* at its ancestral history. A vivid example concerns the cuckoo. To which deplorable bird we now turn.

'Deplorable bird'? Of course I don't really mean that. The phrase amused me in a Victorian bird book belonging to my Cornish grandparents, where it referred to the cormorant. Each page of the book was devoted to one species. When you turned to the cormorant's page, the very first sentence to greet you was, 'There is nothing to be said for this deplorable bird.' I can't remember what grudge the author held against the cormorant. He might have had better grounds with the cuckoo, which is certainly deplorable from the point of view of its foster parents but, as a Darwinian biologist, I think it is a supreme wonder of the world. 'Wonder', yes, but there's also an element of the macabre in the spectacle of a tiny wren devotedly feeding a chick big enough to swallow it whole.

Everyone knows that cuckoos are brood parasites who trick nesting birds of other species into rearing their young. 'Cuckoo in the nest' is proverbial. John Wyndham's *The Midwich Cuckoos*, about aliens implanting their young in unwitting human wombs, is one of several works of fiction whose titles sound the cuckoo *motif*. Then there are cuckoo bees, cuckoo wasps, and cuckoo ants who, in their own hexapod ways, hijack the nurturing instincts of other species of insect. The cuckoo fish, a kind of catfish from Lake Tanganyika, drops its eggs among the eggs of other fish. In this case the hosts are 'mouthbreeders', fish belonging to the Cichlid family who take their eggs and young into their own mouths for protection. The cuckoo fish's eggs and later fry are welcomed into the unsuspecting host's mouth, and tended as lovingly as the mouthbreeder's own.

Plenty of bird species have independently evolved their own versions of the cuckoo habit, for example the cowbirds of the New World, and cuckoo finches of Africa. Within the cuckoo family itself (Cuculidae), 59 of the 141 species parasitise other species' nests, the habit having evolved there three times independently. In this chapter, unless otherwise stated, for the sake of brevity I use the name cuckoo to mean *Cuculus canorus*, the so-called common cuckoo. Alas, it's not common anymore, at least in England. I miss their springtime song even if their victims don't, and was delighted to hear it on a recent visit to a beautiful, remote corner of western Scotland where it 'shouts all day at nothing'. My main authority – indeed today's world authority – is Professor Nick Davies of Cambridge University. His book *Cuckoo* is a delightful amalgam of natural history and memoir of his field research on Wicken Fen, near Cambridge. Described by David Attenborough as one of the country's greatest field naturalists, he achieves heights of lyrical word-painting unsurpassed in the literature of modern natural history:

North towards the horizon is the eleventh-century cathedral of Ely, which sits on the raised land of the Isle of Ely, from where Hereward led his raids against the Normans. In the early mornings, when the mist lies low, the cathedral appears as a great ship, sailing across the fens.

The ruthlessness of the cuckoo begins straight out of the egg. The newly hatched chick has a hollow in the small of the back. Nothing sinister about that, you might think. Until you are told the sole use to which it is put. The cuckoo nestling needs the undivided attention of its foster parents. Rivals for precious food must be disposed of without delay. If it finds itself sharing the nest with either eggs or chicks of the foster species, the hatchling cuckoo fits them neatly into the hollow in its back. It then wriggles backwards up the side of the nest and tosses the competing egg or chick out. There is, of course, no suggestion that it knows what it's doing, or why it is doing it, no feelings of guilt or remorse (or triumph) in the act. The behavioural routine simply runs like clockwork. Natural selection in ancestral generations favoured genes that shaped nervous systems in such a way as to play out this instinctive act of (foster) fratricide. That is all we can say.

And there's no more reason to expect the foster parents to know what they are doing when they fall for the cuckoo's trick. Birds are not little feathered humans, seeing the world through the lens of intelligent cognition. It makes at least as much sense to see the bird as an unconscious automaton. This helps us understand the otherwise surprising behaviour of foster parents. A pioneering cinematographer of the cuckoo's dark ways was Edgar Chance, avid ornithologist of the early twentieth century. By Nick Davies's account of his film, a mother meadow pipit appeared totally unconcerned as it watched its own precious offspring being murdered by the cuckoo chick in its nest. The mother then left on a foraging trip, as if nothing untoward had happened. When she returned, she pointlessly fed her chick as it lay dying on the ground. From a human cognitive point of view, her behaviour makes no sense: neither the impassive watching of the initial murder nor the subsequent futile feeding of the doomed chick. We shall meet this point again and again throughout the chapter.

The name 'cuckoo' is derived from the simple, two-note tune of the male bird's song, so simple indeed that some ornithologists downgrade it from 'song' to 'call' (on parallel grounds to the hysterically unpopular downgrading of Pluto to sub-planet status). The cuckoo's song (or call) is commonly described as dropping through a minor third, but I'm happy to quote no less an authority than Beethoven in support of my hearing it as a major third. His famous cuckoo in the Pastoral Symphony descends from D to B Flat. Whether major or minor, whether song or call, it is simple – and perhaps has to be simple because the male never gets a chance to learn it by imitation. A cuckoo never meets either biological parent. It knows only its foster parents, who could belong to any of a variety of species, each with its own song, which the young cuckoo must not learn. So the male cuckoo's song has to be hard-wired genetically, and a kind of common sense concludes, not very confidently, that it should therefore be simple.

Now we approach the remarkable story that earns the cuckoo its place in a chapter on genes 'looking backwards in time'. Cuckoo eggs mimic the colour and patterning of the other eggs in the particular foster nest in which they sit. And they mimic them even though many different foster species are involved, with very different eggs. Here is a clutch of six brambling eggs plus one cuckoo egg. The only way

I, and doubtless you, can tell which one is the cuckoo egg is by its slightly larger size.

At first sight, such egg mimicry might seem no more remarkable than the 'paintings' of Chapter 2. Well, that's quite remarkable enough! But now look at the next picture showing a parasitised nest of meadow pipit eggs.

Again, you can spot the tell-tale size of the cuckoo egg. But what is really noticeable is that the cuckoo egg in the second picture is dark with black speckles like meadow pipit eggs, whereas the cuckoo egg in the first picture is light and with rusty speckles like brambling eggs. Meadow pipit eggs are dramatically different from brambling eggs. Yet cuckoo eggs achieve a near-perfect colour match in each of the two nests.

Once again, the mimicry might seem par for the course, all of a piece with the lizard, frog, spider, or ptarmigan 'paintings' of Chapter 2. It would indeed be relatively unremarkable if the cuckoos that parasitise bramblings were a different species from the cuckoos that parasitise meadow pipits. But they aren't. They're the same species. Males breed indiscriminately with females reared by any foster species, so the genes of the whole species are mixed up as the generations pass. That mixing is what defines them all as of the same species. Different females, all belonging to the same species and consorting with the same males, parasitise redstarts, robins, dunnocks, wrens, reed warblers, great reed warblers, pied wagtails, and others. But each female parasitises only one of those host species. And the remarkable fact is that (with a few revealing exceptions) the eggs of each female cuckoo faithfully mimic those of the particular host in whose nest she lays them. The only consistent betrayer is that cuckoo eggs are slightly larger than the host eggs that they mimic. Even so, they are smaller than they 'should' be for the size of the cuckoo itself. Presumably, if the pressure to mimic drove them to be any smaller, the chicks would be penalised in some way. The actual size is a compromise between pressure to be small to mimic the host eggs, and an opposite pressure towards the larger optimum for the cuckoo's own size.

I doubt that you're wondering why egg mimicry benefits the cuckoos. Foster parents are mostly very good at spotting cuckoo eggs, and they often eject them. A cuckoo egg of the wrong colour would stand out like a sore thumb. Actually, that's an unusually poor cliché. Have you ever seen a sore thumb, and did it stand out? Let's initiate a new simile. Stands out like a baseball at Lord's? Like a Golden Delicious in a basket of genuinely delicious apples? Just look at that cuckoo egg in the brambling nest and imagine transplanting it into the meadow pipit nest. Or vice versa. The host birds would unhesitatingly toss it out. Or, if tossing it out is too difficult, abandon the nest altogether. Such discrimination is not a surprise when you consider that bird eyes are acute enough to drive the perfecting of the exquisitely detailed painting of lichen-mimicking moths and stick-mimicking caterpillars.

Foster parents, then, whether as automata or cognitively, can be expected to provide the selection pressure that explains why it might benefit cuckoo eggs to show such beautiful egg mimicry. They throw out eggs that don't look like their own. But what is surprising, hugely so, is that cuckoos, all of one intrabreeding species, manage to mimic the eggs of many *different* foster parent species. To drive home the point, here's yet another example: a reed warbler nest with, once again, wonderful egg mimicry by the single, slightly larger cuckoo egg.

These beautiful examples force us back to the central question of this whole discussion. How is it possible for female cuckoos, all belonging to the same species and all fathered by indiscriminate males, to produce eggs that match such a range of very different host eggs? Are we to believe that female cuckoos take one look at the eggs in a nest and take a decision to switch on some kind of alternative egg-colouring mechanism in the lining of the oviduct? That is improbable, to say the least. There are women who might love to control, by sheer willpower and for very different reasons, the behaviour of their own oviduct. But it's not the kind of thing willpower does. And, with the best will in the world, it's not clear how will will power it.

What is the true explanation for the female cuckoo's apparent versatility? Nobody knows for sure, but the best available guess makes use of a peculiarity of bird genetics. As you know, we mammals determine our sex by the XX / XY chromosome system. Every woman has two X-chromosomes in all her body cells, so all her eggs have an

X-chromosome. Every man has an X- and a Y-chromosome in all his body cells. Therefore, half his sperms are Y sperms (and would father a son when coupled with a necessarily X egg) and half are X sperms (would father a daughter when coupled with a necessarily X egg). Less well known is that birds have a similar system, but it evidently arose independently because it is reversed. The chromosomes are called Z and W instead of X and Y, but that's not important. What matters is that in birds females are ZW and males are ZZ. That's opposite to the mammal convention, but otherwise the principle is the same. Whereas the Y-chromosome passes only down the male line in mammals, in birds the W chromosome passes only down the female line. The W comes from the mother, the maternal grandmother, the maternal maternal great grandmother and so on back through an indefinite number of female generations.

Now recall the title of this chapter: 'The Backward Gene's-Eye View'. It's all about genes looking back at their own history. Imagine you are a gene on the W-chromosome of a cuckoo, looking back at your ancestry. Not only are you in a female bird today, you have never been in a male bird. Unlike the other genes on ordinary chromosomes (autosomes), which have found themselves in male and female bodies equally often down the ages, the ancestral environments of the W-chromosome have been entirely confined to female bodies. If genes could remember the bodies they have sat in, the memories of W-chromosomes would be exclusively of female bodies not male ones. Z-chromosomes would have memories of both male and female bodies.

Hold that thought while we look at a more familiar kind of memory: memory by the brain, individual experience. It is a fact that female cuckoos remember the kind of nest in which they were reared, and choose to lay their own eggs in nests of the same foster species. Unlike the improbable feat of controlling your own oviduct, remembering early experience is exactly the kind of thing bird brains are known to do. When they come to choose a mate, as we saw in Chapter 7, birds of many species refer back to a kind of mental photograph of their parent, which they filed away in memory after their first encounter on hatching ('imprinting'): even if – in the case of incubator-hatched goslings, for instance – what they later find attractive is Konrad Lorenz. To remember Lorenz, parental plumage, father's song, or foster-parent's nest – it's all the same kind of problem. The same imprinting brain mechanism works well enough in nature even if, in captivity, it misfires.

I think you can see where this argument is going. Each female mentally imprints on the same foster nest as her mother; and therefore her maternal grandmother; and her maternal maternal great grandmother. And so on back. And her childhood imprinting leads her to choose the same kind of nest as her female forebears. So, she belongs to a *cultural* tradition going exclusively down the female line. Among females there are robin cuckoos, reed warbler cuckoos, dunnock cuckoos, meadow pipit cuckoos, etc., each with their own female tradition. But only females belong to these cultural traditions. Each cultural line of females is called a gens – plural gentes. A female may belong to the meadow pipit gens, or the robin gens, or the reed warbler gens, etc. Males don't belong to any gens. They are descended from – and they father – females of all gentes indiscriminately.

Finally, we put these two strands of thought together, again in the light of the chapter's title. With the exception of W-chromosome genes, all the genes in a female cuckoo look back through a chain of ancestors belonging to every gens that's going. W-chromosomes aside, gentes are not genetically separate like true races, because males confound them. Only W-chromosome genes are gens-specific. Only W-chromosomes look back on ancestors of a particular gens to the exclusion of any other. We talked of two kinds of memory: genetic memory and brain memory. See how the two coincide where W-chromosome genes are concerned!

With respect to the W-chromosome, and only the W-chromosome, gentes are separate genetic races. So – I think you've already completed the argument yourself – if the genes that determine egg coloration and speckling are carried on the W-chromosome, it would solve the riddle we began with, the riddle of how it's possible for the females of one species of cuckoo to mimic the eggs of a wide variety of host species. It isn't willpower that chooses egg colour, it's W-chromosomes.

You will have guessed that it's not as simple as that. Things seldom are in biology. Although female cuckoos have a strong preference for their natal nest type when they come to lay, they occasionally make a mistake and lay in the 'wrong' nest, different from their natal nest. Presumably that's how new gentes get their start. And not all gentes achieve good egg mimicry. Dunnock (hedge sparrow) eggs are a beautiful blue. But cuckoo eggs in dunnock nests aren't blue (left). They aren't even 'trying' to be blue, we might say. The cuckoo egg

in the picture stands out like a sore ... like a bloodhound in a pack of dachshunds. Are cuckoos, perhaps, constitutionally incapable of making blue eggs? No. *Cuculus canorus* in Finland has achieved a most beautiful blue, in perfect mimicry of redstart eggs (right). So why don't cuckoo eggs mimic dunnock eggs? And how do they get away with it? The answer is simple, although it remains puzzling. Dunnocks are among several species that don't discriminate, don't throw out cuckoo eggs. They seem blind to what looks to us glaringly obvious. How is this possible, given that other small songbirds have powers of discrimination acute enough to perfect the finishing touches to the egg mimicry achieved by their respective gentes of female cuckoos? And given that bird eyes are capable of perfecting the detailed mimicry of stick caterpillars, lichen-mimicking moths, and the like?

Cuckoos and their hosts, as with stick caterpillars and their predators, are engaged in an 'evolutionary arms race' with one another.

As mentioned in Chapter 4, arms races are run in evolutionary time. It's a persuasive parallel to human arms races, which are run in 'technological time', and a lot faster. The aerial swerving and dodging chases of Spitfires and Messerschmitts were run in real time measured in split seconds. But in the background and more slowly, in factories and drawing-offices in Britain and Germany, races were run to improve their engines, propellers, wings, tails, weaponry, etc., often in response to improvements on the other side. Such technological arms races are run over a timescale measured in months or years. The arms races between cuckoos and their various host species have been running for thousands of years, again with improvements on each side calling forth retaliatory improvements in the other.

Nick Davies and his colleague Michael Brooke suggest that some gentes have been running their respective arms races for longer than others. Those against meadow pipits and reed warblers are ancient arms races, which is why both sides have become so good at outdoing the other – and therefore why the cuckoo eggs are such good mimics. The arms race against dunnocks, they suggest, has only just begun. Not enough time for the dunnocks to evolve discrimination and rejection. And not enough time for the dunnock gens of cuckoos to evolve the appropriate blue colour.

If it's true that cuckoos have only just 'moved into' dunnock nests, we must suppose that these 'pioneer' cuckoos have 'migrated' from another host species, presumably one with rusty-spotted grey eggs because that's the egg colour of the 'newly arrived' dunnock gens of cuckoo. I suppose this is how any new gens gets its start. But don't be misled by 'pioneer' and 'migrated'. It would not have been any kind of bold decision to sally forth into fresh nests and pastures new. It would have been a mistake. As we've seen, cuckoos do indeed occasionally lay an egg in the wrong kind of nest, a nest appropriate to a different gens. Their egg then really does stand out like a ... invent your own substitute for the sore thumb cliché. Natural selection normally penalises such blunders, we can presume, pretty promptly. But what if it's a new host species that hasn't yet been 'invaded' by cuckoos. The new host species is naive. They haven't hitherto had any reason to throw out mismatched eggs. Once again, remember, birds are not little feathered humans with human judgement. The arms race has yet to get properly under way. And the host species can expect to remain naive while the arms race is yet young. But how young is young? Strangely enough, we are not totally without evidence bearing on the question, as Nick Davies points out.

Call the witness Geoffrey Chaucer. In *The Parlement of Foules* (1382), the cuckoo is reproached: 'Thou mordrer of the heysugge on the braunche that broghte thee forth.' Another name for dunnock is hedge sparrow or, in Middle English, heysugge (heysoge, heysoke, eysoge). This would seem to suggest that cuckoos were already parasitising dunnocks in the fourteenth century, when Chaucer wrote. Is 650 years long enough for an arms race to reach some sort of perfection of mimicry? Perhaps not, given that, as Davies points out, only 2 per cent of dunnock nests are parasitised. Maybe, then, the selection pressure is so weak that a 600-year-old arms race is indeed young.

I prefer to add two further suggestions. The first concerns identification. Did Chaucer really mean dunnock when he said heysugge? When we say 'sparrow' we normally mean the house sparrow, *Passer domesticus*, not the hedge sparrow or dunnock, *Prunella modularis*. Yet the English word 'sparrow' is used for both. To many who are not avid twitchers, all little brown birds (LBBs) look much the same, and we might even sink so low as to call them all 'sparrows'. I can't help wondering whether Chaucer was using 'heysugge' to mean LBB rather than specifically *Prunella modularis*?

My second suggestion is more biologically interesting. If we think carefully about it, there's no reason, is there, to suppose that there's only one cuckoo gens for each host species? Maybe Chaucer's gens of dunnock cuckoos has died out, and a new gens of dunnock cuckoos is just beginning its arms race. Perhaps other gentes of dunnock cuckoos have perfect egg mimicry today, but have not come to the notice of ornithologists. There would be no relevant gene flow between them because males don't have W-chromosomes.

Claire Spottiswoode and her colleagues are running a parallel study of an unrelated South African finch, which convergently evolved the cuckoo habit. The cuckoo finch, *Anomalospiza imberbis*, lays its eggs in the nests of grass warblers. Different gentes of cuckoo finch mimic the eggs of different grass warbler species. There is genetic evidence that what distinguishes the gentes is indeed their W-chromosomes, which reinforces the idea that the same thing is going on in cuckoos. As Dr Spottiswoode points out, this doesn't have to mean that every detail of all the egg colours is carried on the W-chromosome. In both cuckoos and cuckoo finches, genes for making all the different egg colours have very probably been built up on other chromosomes ('autosomes') over many generations, and are carried by all the gentes and passed on by males as well as females. The W-chromosome need only have switch-genes – genes that switch on or off whole suites of genes carried on autosomes. And the relevant autosomal genes would be carried by males as well as females.

This is indeed how sex itself is determined. If you have a Y-chromosome, you have a penis. If you have no Y-chromosome, you have a clitoris instead. But there's no reason to suppose that the genes that influence the shape and size of a penis are confined to the Y-chromosome. Far from it. It's entirely plausible that they are scattered over many autosomes. There's no reason to doubt that a man may inherit genes for penis size from his mother as well as from his father. Presence or absence of a Y-chromosome determines only which alternative suite of genes on autosomes will be switched on. For most purposes you can think of the entire Y-chromosome as a single gene that switches on suites of other genes on autosomes elsewhere in the genome. A point of terminology: members of these suites of autosomal genes are called 'sex-limited' as distinct from 'sex-linked'. Sex-linked genes are those that are actually carried on the sex-chromosomes themselves.

Probably the best guess towards a solution of the riddle of cuckoo egg mimicry is that suites of genes on lots of chromosomes determine egg coloration and spotting. These are equivalent to 'sex-limited', and we may call them 'gens-limited'. They are switched on or off by the presence or absence of one or more genes on the W-chromosome, genes that, by analogy, we can call 'gens-linked'. All cuckoo autosomes may have suites of genes for mimicking a whole repertoire of host eggs. W-chromosomes contain switch genes that determine which suite of genes is turned on. And it is W-chromosomes that are peculiar to each gens of females, W-chromosomes that look back at their history and see a long line of nests of only one foster species.

This interpretation of egg mimicry in cuckoos is my introduction to the topic of the backward gene's-eye view, genes looking over their shoulder at their own ancestry. Here's a similar but more complicated example involving fish and the Y-chromosome. Different kinds of fish display a bewildering variety of sex-determining systems. Some don't use sex chromosomes at all but determine sex by external cues. Some fish are like birds in that females are XY and males are XX. Others are like us mammals: males are XY, females XX. Among these are small fish of the genus *Poecilia*, which includes mollies and guppies among popular aquarium fish. One species, *Poecilia parae*, has a remarkable colour *polymorphism*, which affects only the males. Polymorphism means that there are different genetically determined colour types coexisting in the population (in this case five colour patterns) and the proportions of the different types remain stable in the population through time. All five male morphs can be found swimming together in South American streams. There's only one female morph: females look alike.

Since the polymorphism affects only one sex, we can call them five gentes, by analogy with the cuckoos, with the difference that in these fish it's the males who are separated by gens. The picture shows the five male types plus a female at the bottom. Three of the five male types have two long stripes like tramlines. Between the tramlines there is colour, and I'll call them reds, yellows, and blues respectively. These three 'tramliners' can, for many purposes, be lumped together. The fourth type has vertical stripes. They're officially named 'parae', but confusingly that's also the name of the whole species. I'll call them 'tigers'. The fifth type, 'immaculata', is relatively plain grey, like females but smaller, and I'll call them 'greys'.

Tigers are the largest. They behave aggressively, chasing rival males away, and copulating with females by force. Greys are the

Tiger

Grey

Blue

Yellow

Red

Female

Male 'gentes' in fish?

smallest, and they manage to copulate only by occasionally sneaking up on females opportunistically. When they get away with it, it seems to be because otherwise aggressive males mistake them for females, which they do indeed resemble. Greys have the largest testes, presumably capable of producing the most sperm, perhaps to take advantage of their scarce opportunities to use it. Red, yellow, and blue tramliners are of intermediate size. Rather than rape or sneak, they court females in a civilised manner, displaying their respective coloured flanks.

Now here's where the parallel to cuckoos kicks in. Evidence suggests that colour morph inheritance runs entirely down the male line. In every case studied, sons belong to the same type as their father, and therefore paternal grandfather, paternal paternal great grandfather, etc. Their mother has no genetic say in the matter, and nor does their maternal grandfather, etc., even though each one belongs to one colour gens or another. This suggests the hypothesis that the five types of males differ with respect to their Y-chromosomes – just as gens-inheritance in female cuckoos seems to be carried on the W-chromosome. The details of colour pattern and behaviour of the male fish may be carried in suites of genes on autosomes (gens-limited). But the genes determining which gens an individual belongs to (and presumably which suite of colour and pattern genes on other chromosomes is switched on) seem to be gens-linked, that is, carried on the Y-chromosome.

Researchers are doing fascinating work on mate choice in these fish and are homing in on what maintains the polymorphism. It seems that each of the five male types has an equilibrium frequency, fitting the definition of a true polymorphism. If its frequency falls below the equilibrium, it is favoured and therefore becomes more frequent in the population. If its frequency rises too high, it is penalised and its frequency decreases. This so-called 'frequency-dependent selection' is a known way for polymorphisms to be maintained in a population. How might it work in practice? The details are not yet clear but might look something like this. The grey sneakers benefit from being mistaken for females. If they become too frequent, perhaps the real females or aggressive tigers get 'wise' to them. How about the tigers themselves? If they get too frequent, they waste time fighting each other instead of mating. This might give the greys more opportunity to sneak matings. As for the three 'tramliners', who court females in a gentlemanly manner by flashing their vividly coloured flanks, there is some evidence that females prefer rarer types. This would fit the 'equilibrium frequency' idea, although it's not clear why females should exhibit such a preference. More research is needed and is under way now. I am grateful to Dr Ben Sandkam, formerly of the University of British Columbia and now at Cornell, for sharing with me his thoughts on these matters.

Now let's again apply the backward-looking technique of this chapter. Every male of *Poecilia parae* can look back through a long line of male ancestors, all belonging to the same gens as him, and all sharing the same Y-chromosome. This is what makes it possible for suites of genes for colour patterning and associated behaviour to become switched on in separate gentes of males, despite their sharing the same ancestors in the female line. The gene's-eye view of the past comes into its own again, as with the cuckoos. Autosomal genes, governing characteristics other than gens-specific colour, look back on ancestors of all gentes.

Returning to cuckoos, the 'looking back' ploy can help us answer another riddle, and it's an even tougher one. Although most host species are very good at distinguishing cuckoo eggs from their own (how else could natural selection have perfected cuckoo egg mimicry?), they turn out to be lamentably bad later, failing to notice that the growing cuckoo fledgling is an impostor. Even though it dwarfs them, in most cases grotesquely so. A tiny warbler is in danger, you might think, of being swallowed whole by its monstrous foster child. Foster parents, of whatever species, end up dwarfed by the cuckoo nestling into whom they tirelessly shovel food, working every devoted daylight hour to do so. How do the cuckoo nestlings get away with such a transparent, over-the-top deception? Once again, we have to be

A warbler feeding a cuckoo

more than usually on our guard against anthropomorphism. Do not ask whether the bird's behaviour makes sense from a human-like cognitive perspective. Of course it doesn't. Ask instead about selection pressures acting on ancestral genes that control the development of behavioural automatisms.

Even given this preliminary, I must admit that available answers to the riddle epitomised by the picture on the previous page remain unsatisfying, compared to the explanations that I am accustomed to offering in my books. And indeed, compared to the explanation of egg mimicry. But here's the best explanation – or series of partial explanations – I can find. We return to the idea of the arms race. In our 1979 paper, John Krebs and I considered ways in which an arms race might end in 'victory' for one side (here again, the quotation marks are strongly advised). We identified two principles, the 'Life Dinner' and the 'Rare Enemy' principle. These are closely related, maybe just different aspects of the same thing.

In one of Aesop's Fables, a hound was pursuing a hare, got tired and gave up. Taunted for his lack of stamina, the hound replied, 'It's all very well for you to laugh, but we had not the same stake at hazard. He was running for his life, but I was only running for my dinner.'

As in military arms races, predators and prey must balance design improvements and resources against economic costs. The more they put into servicing the arms race – muscles, lungs, heart, the machinery of speed and endurance – the less is available for other aspects of life such as making eggs or milk, building up fat reserves for the winter etc. In the language of Darwinism, Aesop's hares have been subject to stronger selection to invest resources into the arms race than the hounds. There is an asymmetry in the cost of failure – loss of life versus mere loss of dinner. The failed predator lives to pursue another prey. The failed prey has fled its last pursuer. But now, notice how we can say the same thing more piercingly in the language of the genetic book of the dead. The predator's genes can look back on ancestors many of whom were outrun by prey. But not one of the prey's ancestors was outrun by a predator. At least not before it had passed on its genes. Plenty of predator genes can look back on ancestors who failed to outrun prey. Not a single prey gene can look back on ancestors who had lost a race against a predator.

Apply the Life Dinner Principle to the cuckoo nestling and its host. The cuckoo nestling can look back on an unbroken line of ancestors, literally not a single one of whom was outwitted by a discriminating host. If it had been, it would not have become an ancestor. Cuckoo genes for failing to fool hosts are never passed on. But genes that lead foster parents to fail to notice cuckoos? Plenty of hosts who were fooled by cuckoos could live to breed again. Genetic tendencies among hosts to be fooled by cuckoos can be passed on. Genetic tendencies among cuckoos to fail to fool hosts are never passed on. It's the Life Dinner Principle in operation.

Moreover, the host can look back on ancestors many of whom may never have met a cuckoo in their lives. In Nick Davies and Michael Brooke's long-running study on Wicken Fen, only 5 to 10 per cent of reed warbler nests were parasitised by cuckoos. And this brings us to the Rare Enemy Effect. Cuckoos are comparatively rare. Most reed warblers, wagtails, pipits, dunnocks, etc. probably get through their lives and successfully reproduce without ever encountering a cuckoo. They may look back on many ancestors who never encountered a cuckoo in their lives. But every single cuckoo looks back at an unbroken line of ancestors who successfully fooled a host into feeding them. Asymmetries of this kind could favour 'victory' such that even a monstrous cuckoo nestling gets away with fooling its diminutive foster parent. The selection pressure to outwit cuckoos is weak compared to the selection pressure on cuckoos to do the outwitting.

Another parable with an Aesopian flavour is the fable of the boiled frog. A frog dropped into very hot water might do anything in its power to jump out. But a frog in cold water that is slowly heated up does not notice until it is too late. When the baby cuckoo first hatches, the deceiver is indistinguishable from the real thing. As it gradually grows, there is no one day when it suddenly becomes obvious that it's a fake. Just as there's never a day when a baby becomes a child; or a child a teenager; or a middle-aged man old. Every day, it looks much the same as the day before. Perhaps this helps the outwitting.

Note that the boiled frog effect doesn't apply to eggs. A cuckoo egg suddenly appears in the nest. It doesn't gradually become more and more imposterish like a cuckoo nestling.

In another pair of papers already mentioned, Krebs and I proposed that animal communication in general can be seen as *manipulation*. I discussed this in Chapter 7 in connection with nightingale song bewitching John Keats. Birdsong is known to cause female gonads to swell. This is an example of what we called manipulation. It will not always be to the female's advantage to submit to it. There will be an arms race between salesmanship and sales-resistance, each side escalating in response to the other. What tricks of salesmanship might the cuckoo nestling employ, in response to the sales-resistance of the host? They'd need to be pretty powerful to outweigh the eventually incongruous mismatch in size between foster parent and cuckoo nestling. But that's no argument against their existence.

All nestlings open their gapes wide and squawk their appeals for food. If you're a baby reed warbler, say, the louder you cry, the more likely you are to persuade your parent to drop food into your gape rather than a sibling's (and there is indeed good Darwinian reason for competition among siblings, even real gene-sharing siblings). On the other hand, loud vocalisation costs vital energy. This applies to baby birds as much as to adults. In one study of wrens at Oxford, the researcher allowed himself to speculate that a male literally sang himself to death. The calling rate and loudness of a baby reed warbler will normally be regulated to an optimum level: enough to compete with siblings, but not so much as to overtax itself or attract predators. The oversized baby cuckoo needs as much food as four young reed warblers. It urges the foster parent on by sounding like a clutch of reed warbler chicks rather than just one very loud reed warbler chick.

Among the ingenious field experiments done by Nick Davies, he and his colleague Rebecca Kilner put a blackbird nestling in a reed warbler nest. The young blackbird was about the same size as a cuckoo nestling. The reed warblers fed it, but at a lower rate than they would normally feed a baby cuckoo. Then the experimenters played their masterstroke: a sound recording of a baby cuckoo piped through a little loudspeaker next to the nest, switched on whenever the baby blackbird was seen to beg. Now the reed warbler adults upped the rate with which they fed the blackbird chick, to a rate appropriate to a baby cuckoo – the same rate as for a clutch of baby reed warblers. And indeed, a recording of four baby reed warblers crying had the same effect. It would seem that baby cuckoo squawks have evolved to become a super-stimulus. Super-stimuli are well attested in experiments on bird behaviour. My old maestro Niko Tinbergen reported that oystercatchers, offered a choice, will preferentially attempt to incubate a dummy egg eight times the volume of their own egg. It's called a *supernormal* stimulus. Something like this is what we'd expect as the culmination of an evolutionary arms race, with escalating salesmanship on the cuckoo's side keeping pace with escalating sales-resistance on the part of the foster parents.

How about a visual equivalent of such a super-stimulus? The open beak of all nestlings is conspicuous, often bright yellow, orange, or red. Doubtless such bright coloration persuades the parents to drop food in, the brighter the gape the greater the chance of their favouring this gape rather than a sibling's. Reed warbler chicks have a yellow gape. Davies and colleagues found that reed warbler parents gauge their food-fetching efforts according to the total area of yellowness gaping at them in the nest, and also to the rate of begging cries. Cuckoo chicks have a red gape. Is this, perhaps, a stronger stimulus than yellow? An experiment with painted gapes failed to support the hypothesis. Is the cuckoo gape, then, larger than a reed warbler chick's gape? Yes, cuckoo chicks have a bigger gape than any one reed warbler chick. But its area is not equal to the sum of four reed warbler chicks – perhaps closer to two. Cuckoo chicks use sound to compensate for this, and by two weeks of age a cuckoo chick sounds like a clutch of reed warbler chicks. The combination of a somewhat bigger gape than one reed warbler chick's, together with supernormal begging cries, is just enough to persuade the adult reed warblers to pump into the cuckoo chick as much food as they would normally bring to a whole clutch of their own chicks. Once again, we could see the supernormal begging call as the end product of an escalating arms race between salesmanship and sales-resistance.

That birds are susceptible to large gapes – even the alien gape of a fish – is shown by the well-attested observation of a cardinal (an American bird) repeatedly dropping food into the open mouth of a goldfish. We view the scene through human eyes and think, how absurd, how could a bird be so stupid? But the example of the oystercatcher sitting on the giant egg should warn us that human eyes are precisely what we should not trust. We have no right to be sarcastic. Birds are not little humans, cognitively aware of what they are doing and why they are doing it. And after all, a human male can be sexually aroused by a supernormal caricature of a female, even though he is well aware that it is a drawing on two-dimensional paper, with unnaturally exaggerated features, and a fraction of normal size. The baby cuckoo has no idea what it is doing when it tosses eggs out of the nest. Think of it as a programmed automaton. The oystercatcher does not know why it sits on a giant egg. Think of it as a pre-programmed incubation machine. And in the same way, think of a parent bird as a robot mother, programmed to drop food into wide-open gapes, however ridiculous it may seem to us when the gape belongs to a fish. Or to the giant imposter who is a nestling cuckoo.

If cuckoo nestlings have a supernormal gape, mimicking two ordinary chicks, there's an Asian cuckoo, Horsfield's hawk cuckoo, *Cuculus fugax*, that goes one better. It has the visual equivalent of a clutch of gapes. In addition to its yellow gape, it has a pair of dummy gapes: a patch of bare skin on each wing, the same yellow colour as the real gape. It waves the wing patches about, usually one at a time, next to the real gape. The foster parent (a species of blue robin was the host in this Japanese study by Dr Keita Tanaka) is stimulated by the double whammy of gape plus patch. Dr Tanaka has kindly sent me several photographs plus some amazing film footage. As soon as the foster parent flies in, the cuckoo chick dramatically raises its right wing and waves it about. The gesture reminds me of a swordsman raising his shield to intercept an attack. But this analogy has it exactly

A cardinal feeding a goldfish

wrong. The point is not to repel but to attract. One film even shows the robin vigorously stuffing food up against the yellow patch on the upheld right wing, before turning and shoving it into the wide-open gape instead. The Japanese researchers ingeniously blacked out the wing patch, and this reduced the feeding rate by the robins. There's a similar story for another brood parasite, the whistling hawk cuckoo, *Hierococcyx nisicolor*, in China. Like the Horsfield's hawk cuckoo, the nestlings have yellow wing patches that they display in the same way, to fool foster parents.

So much for cuckoos, not deplorable because a true wonder of nature and natural selection. Now, let's see what else we can do with the notion of genes looking over their shoulder.

Horsfield's hawk cuckoo with fake gape on wing

11

More Glances in the
Rear-View Mirror

Where once they would have talked of the good of the species, nowadays essentially all serious biologists studying animal behaviour in the wild have adopted what I am calling the gene's-eye view. Whatever the animal is doing, the question these modern workers ask is, 'How does the behaviour benefit the self-interested genes that programmed it?' David Haig, now at Harvard University, is one of those pushing this way of thinking towards the limit, illuminating a great diversity of topics, including some important ones that doctors should care about, such as problems of pregnancy.

Among other things, Haig noticed a lovely example of genes looking backwards – actually at the immediate past generation. There's a phenomenon called genomic imprinting. A gene can 'know' (by a chemical marker) whether it came from the individual's father or mother. As you can imagine, this radically changes the 'strategic calculations' whereby a gene looks after its own self-interest. Haig shows how genomic imprinting changes how a gene views kin. Normally, a gene for kin altruism should regard a half-sibling as equivalent to a nephew or niece – half the value of a full sib or offspring. But if the altruistic gene 'knows' it came from the mother and not the father, it should see a maternal half-sibling as equal to its own offspring, or to a normal full sibling. The other way round if it 'knows' it came from the father. It should then see the maternal half-sibling as equivalent to an unrelated individual. Genomic imprinting opens up a whole lot of ways in which genes within an individual can come into conflict with one another, the topic of Burt and Trivers' book *Genes in Conflict*. Haig goes so far as to blame warring genes for the familiar psychological sensation of being pulled in two directions at once, as in short-term gratification versus longer-term benefit. Genomic imprinting provides a stark example of how a gene might look in the 'rear-view mirror'. Other examples constitute the topics of this chapter.

A gene on a mammalian Y-chromosome 'looks back' at an immensely long string of ancestral male bodies and not a single female one, probably as far back as the dawn of mammals if not further. Our mammal Y-chromosome has been swimming in testosterone for perhaps 200 million years. But if Y-chromosomes look back at only male bodies, what about X-chromosomes? If you are a gene on an X-chromosome, you might come from the animal's father, but you are twice as likely to come from its mother. Two-thirds of your ancestral history has been in female bodies, one-third in male bodies. If you are a gene on a chromosome other than a sex chromosome, an autosome, half your ancestral history was in female bodies, half in male bodies. We should expect many autosomal genes to have sex-limited effects, programmed with an IF statement: one effect whenever they find themselves in a male body, a different effect when in a female body.

But when any gene looks back at the male bodies that it has inhabited, what it sees will not be a random sample of male bodies but a restricted sub-set. This is because the average male is often denied the Darwinian privilege of reproduction. A minority of males monopolises the mating opportunities. Most females, on the other hand, enjoy close to the average reproductive success. Red deer stags with large antlers prevail in fights over access to females. So when a red deer gene looks back at its male ancestors, it will see the minority of male bodies that are topped by abnormally large antlers.

Even more extreme is the asymmetry shown by seals, especially *Mirounga*, the elephant seal. There are two species: the southern elephant seal, which I have seen, close enough to touch (though I would not), on the remote island of South Georgia, and the northern elephant seal, which Burney Le Boeuf has thoroughly studied on the Pacific beaches of California. Like many mammals, elephant seals have harem-based societies but they carry it to an extreme. Successful males, 'beachmasters', are gigantic: up to 4 metres long and weighing 2 tonnes. Females are relatively small and are gathered into harems, which may typically number as many as fifty 'belonging to', and vigorously defended by, a single dominant male. Most of the males in the population have no harem, and either never reproduce or bide their time hoping to sneak an occasional copulation, as well as aspiring eventually to get big and strong enough to displace a beachmaster. In one report from Le Boeuf's long-term California study of northern elephant seals, only eight males inseminated an astonishing 348 females. One male inseminated 121 females, while the great majority of males had no reproductive success at all. An elephant seal gene on a Y-chromosome looks back at, not just a long sequence of male bodies, but specifically at the overgrown, blubbery, belching, bloated bodies of a tiny minority of dominant, harem-holding beachmasters: highly aggressive males, over-endowed with testosterone and with the dangling trunks used as living trombones to resonate roars that intimidate other males. On the other hand, an elephant seal gene will look back at a succession of female bodies that are close to the average.

Do you find something puzzling about the fact that only a small minority of males does almost all the fathering? Isn't it terribly wasteful? Think of all those bachelor males, consuming a fat slice of the food resources available to the species, yet never reproducing. A 'top-down' economic planner with species welfare in mind would protest

Sexual inequality on the beach

that most of those males shouldn't be there. Why doesn't the species evolve a skewed sex ratio such that only a few males are born: just enough males to service the females, the same number of males as would normally hold harems? They wouldn't have to fight each other, they'd all get a harem as a matter of automatic entitlement, just for being male. Wouldn't a species with such an economically sensible, planned economy prevail over the present, wildly uneconomical, strife-ridden species? Wouldn't the planned economy species win out in natural selection?

Yes, if natural selection chose between species. But, contrary to a widespread misunderstanding, it doesn't. Natural selection chooses between genes, by virtue of their influence on individuals. And that makes all the difference. If the sensible planned economy were to come about by Darwinian means, it would have to be through the natural selection of genes controlling the sex ratio. This is not impossible. A gene could bias the number of X sperms versus Y sperms produced by males. Or it could favour selective abortion of some male foetuses. Or it could favour starving some baby sons to death and keeping just a favoured few. Never mind how it does it, just call this hypothetical gene the Planned Economy Gene, pegged to top-down common sense.

Imagine a planned economy population where most of the individuals are female, say one male for every ten females. This is the kind of population our sensible economist would expect to see. It is economically sensible because food is not wasted on males who are never going to reproduce. Now imagine a mutant gene arising, a mutation that biases individuals towards having sons. Will this male-favouring gene spread through the population? Alas for the planned economy, it certainly will. In the planned economy, females outnumber males ten to one, so a typical male can expect ten times as many descendants as a typical female. It's a bonanza for males. The son-biased mutant gene will spread rapidly through the population. And the males will have good reason to fight. It's the flip side of our observation that our hypothetical gene looks back at a successful minority of male bodies, not at an average sample of male bodies.

Will the population sex ratio swing right round to the opposite extreme and become male-biased? No, natural selection will stabilise the sex ratio we actually see, a 50/50 sex ratio (but see the important reservation below) with a minority of harem-holding males and a majority of frustrated bachelors. Here's why. If you have a son, there's a good chance he'll end up a disconsolate bachelor who'll give you no grandchildren. But if your son does end up a harem-holder, you've hit the jackpot where grandchildren are concerned. The expected reproductive success of a son, averaged over his slim chance of the jackpot plus the much greater chance of bachelor misery, equals the expected average reproductive success of a daughter. Equal sex ratio genes prevail, even though the society they create is so horribly uneconomical. Sensible as it sounds, the 'planned economy' cannot be favoured by natural selection. In this respect at least, natural selection is not a 'sensible' economist.

I said that selection would stabilise the sex ratio at 50/50 but I added a cautionary reservation. There are various reasons for that caution, and they are important. Here's one of them. Suppose it costs twice as much to rear a son as to rear a daughter. To equip a son to fight off rivals and win a harem, he must be big. Being big doesn't come free. It costs food. If a mother seal must suckle a son for longer than a daughter, if a son costs twice as much as a daughter to rear, the 'choice' facing the mother is not 'Shall I have a son or a daughter' but 'Shall I have a son or two daughters?' The general principle, first clearly understood by RA Fisher, is that the sex ratio stabilised by natural selection is 50/50 *measured in economic expenditure* on daughters versus economic expenditure on sons. That will amount to 50/50 in numbers of male and female bodies, only if the cost of making sons and daughters is the same. Fisher's principle balances what he called *parental expenditure* on sons versus daughters. This may cash out in the form of equal numbers of males and females in the population, but only if sons and daughters are equally costly to rear. There are other complications, some pointed out by WD Hamilton, but I won't stay to deal with them.

Elephant seals are an extreme example of a principle that typifies many mammal species. Females tend to have nearly the same reproductive success as each other, close to the population average, while a minority of males enjoys a disproportionate monopoly of reproduction. In statistical language, *mean* reproductive success of males and females is equal, but males tend to have a higher *variance* in reproductive success. And, to return to the title of this chapter, the ancestral females that genes 'look back on' will be close to the average. But they'll look back on an ancestral history dominated by a minority of males: that minority endowed with whatever it takes in the species concerned – large antlers, fearsome canine teeth, sheer bodily bulk, courage, or whatever it might be.

'Courage' can be given a more precise meaning. Any animal must balance the short-term value of reproducing now against its own long-term survival to reproduce in the future. A brutal fight against a rival male may end in victory and a harem. But it may end in death, or serious injury which presages death. Courage is at a premium. Risking death is worthwhile because the stakes for a male are so high: a huge number of pups to his name if he wins, zero and perhaps death if he loses. A female seal would give higher priority to surviving to reproduce next year. She only has one pup in a year, so she'll maximise her reproductive success by surviving herself. Natural selection would favour females who are more risk-averse than males; would favour males who are more courageous or foolhardy. Males are biased towards a high-stakes high-risk strategy. This is probably why males tend to die younger. Even if they're not killed in battle, their whole physiology is skewed towards living to the full while young, even at the expense of living on at all when old.

A complication is that, in some species, including elephant seals, subordinate males sneak surreptitious matings at the risk of punishment from dominant males. They may adopt a particular strategy known as the 'sneaky male' strategy. This means that as a Y-chromosome looks back at its history, it will see mostly a river of dominant harem-holders but also a side rivulet, that of the sneaky males. And now, a change of topic.

As will be apparent by now, my late colleague WD Hamilton had a restless and highly original curiosity, which led him to solve many outstanding riddles in evolutionary theory, problems that lesser intellects never even recognised as problems. A naturalist from boyhood, he noticed that many insect species come in two distinct types which could be named 'dispersers' and 'stay-at-homes'. Dispersers typically have wings. 'Stay-at-homes' often don't. It's surprising how many species of insects have both winged and wingless members, seemingly in balanced proportion. If you like human parallels, think of human families in which one brother comfortably inherits the farm while the other brother emigrates to the far side of the world in search of an improbable fortune. In the case of plants, dandelion seeds with their fluffy parachutes are 'winged' dispersers, while other members of the daisy family have, to quote Hamilton, 'a mixture of winged and wingless within a single flower head'.

To stolid common sense, it seems intuitively obvious that if parents live in a good place (and they probably do live in a good place, or they wouldn't have succeeded in becoming parents), the best strategy for an offspring must be to stay in the same good place. 'Stay at home and mind the family farm' would seem to be the watchword, and that was the conventional wisdom among most evolutionary theorists before Bill Hamilton. Bill suspected, by contrast, that selection would favour a balance between stay-at-homes and dispersers, the point of balance varying from species to species. He enlisted the help of his mathematical colleague Robert May, and together they developed mathematical models that supported his intuition.

My own, less mathematical way to express Bill's intuition is in terms of the gene's-eye view of the past. No matter how favourable the 'family farm' – the environment in which parents have flourished – it is sooner or later going to be subject to a catastrophe: a forest fire perhaps, or a disastrous flood or drought. So, as a gene looks back at the history of 'the family farm', the parental, grandparental, and great grandparental generations may indeed have flourished there. The success story might go back an unbroken ten or even twenty generations. But eventually, if it looks far enough back into the past, the stay-at-home gene will eventually hit one of those catastrophes.

The disperser gene may look back on the recent past as one of comparative failure: life on the family farm was milk and honey. But if we look back sufficiently far, we come to a generation where only the disperser gene, the gene for wild wanderlust, made it through. There's also the anthropomorphic point that wanderlust occasionally strikes gold.

Naked mole rat

I perhaps went too far when in 1989 I published a speculation about naked mole rats, but it serves to dramatise the point. Naked mole rats are small, spectacularly ugly (by human aesthetics) African mammals, who live underground. They are famous among biologists as the nearest mammalian approach to social insects: ants and termites. They live in large colonies of as many as 100 individuals in which only one female, the 'queen', normally reproduces, and she is fecund enough to compensate for the near sterility of all the other females, who function as 'workers'. A colony can extend through a huge network of 2 or 3 miles of burrows, gathering underground tubers as food.

This much has become lore among biologists intrigued by the obvious similarity to social insects. However, one discrepancy always worried me. Although the ants and termites that we ordinarily see are wingless, sterile workers, their underground nests periodically erupt in a boiling mass of winged reproductive individuals of both sexes. These fly up to mate, after which the newly fertilised young queens settle down, lose their wings (in many cases even biting them off), dig a hole, and attempt to found a new underground nest with the aid of sterile, wingless worker daughters (and sons in the case of termites). The winged castes are Hamilton's dispersers, and they are an essential part – indeed, the essential part – of the biology of social insects. You could say they are what the whole social insect enterprise is all about. Why don't naked mole rats have an equivalent? Their lack of a dispersal phase is something approaching a scandal!

Not literally winged dispersers! Even I am not foolhardy enough to predict rodents with wings. But I did wonder, and still do, whether there might be a dispersal phase that nobody has spotted yet. In 1989 I wrote: 'Is it conceivable that some already known hairy rodent, running energetically above ground and hitherto classified as a different species, might turn out to be the lost caste of the naked mole rat?' My idea for a hitherto unrecognised dispersal caste may not have much going for it, but it is at least testable, a virtue that scientists value highly. The genome of the naked mole rat has been sequenced. If my hypothetical dispersal phase were ever discovered, some hairy mole rats should turn out to have the same genes.

I admitted the implausibility of my suggestion. How could such a hypothetical creature have been overlooked by biologists? However, I went on to make a comparison with locusts. Locusts are the terrifying 'wanderlust' phase of harmless 'stay-at-home' grasshoppers. They look different from grasshoppers and behave very differently. They are the very same grasshoppers but (*oh, in a moment*) they change. The genes of a harmless grasshopper have the capacity, when the conditions are right, to change (*change utterly, and a terrible beauty is born*). The devastating effects are all too well known. My point is that locust plagues only occasionally happen. It just takes the right conditions. Perhaps the dispersal phase of the naked mole rat has yet to erupt during the decades since biologists have been around to study the species? No wonder it has never yet been seen. Perhaps it would take only a crafty hormone injection … and a naked mole rat could become its own hairy, scurrying (though not, I suppose, winged) dispersal phase.

Another change of topic before we leave the backwards gene's-eye view. There are two ways in which we can look back at a family tree. Conventional pedigrees trace ancestry via individuals. Who begat whom? Which individual was born of which mother? The most recent individual ancestor shared by the late Queen Elizabeth II and her husband Prince Philip was Queen Victoria. But you can also trace the ancestry of a particular gene, and you will have guessed that this is the alternative manner of tale I want to tell here. Genes, like individuals, have parent genes and offspring genes. Genes, as well as individuals, have pedigrees, family trees. But there is a significant difference between a 'people tree' and a 'gene tree'. An individual person has two parents, four grandparents, eight great grandparents, etc. So a people tree is a vast ramification as you look backwards in time. Any attempt to draw it out completely will soon get out of hand. The best way to visualise it is not on paper but zooming around a computer screen. Not so the gene tree. A gene has only one parent, one grandparent, one great grandparent, etc. A gene tree is therefore a simple linear array streaking back in time, whereas a people tree bifurcates its way unmanageably into the past. This is not so when you look forwards in time, by the way. A gene can have many offspring but only ever one parent. Looking forwards, gene trees branch and branch. But this chapter is all about looking backwards.

A particular sub-lethal gene, haemophilia, has plagued the royal families of Europe ever since the early nineteenth century. The gene tree of royal haemophilia is simple and fits the page comfortably. The equivalent people tree would want several square metres of paper to be legible. The royal haemophilia gene can be traced back to a particular individual ancestor, Queen Victoria, one of whose two X-chromosomes bore the gene. The mutation occurred, to quote Steve Jones's mordant phrase, 'in the august testicles' of her father, Edward, Duke of Kent. One of Victoria's four sons, Prince Leopold, suffered from haemophilia. The other sons, including Edward VII and his descendants such as our present monarch, King Charles III, beat the odds and were lucky to escape. Leopold survived to the age of thirty, long enough to have a daughter, Princess Alice of Albany, who inevitably carried the gene on one of her X-chromosomes. Her son Prince Rupert of Teck realised his 50 per cent probability of being afflicted and died young.

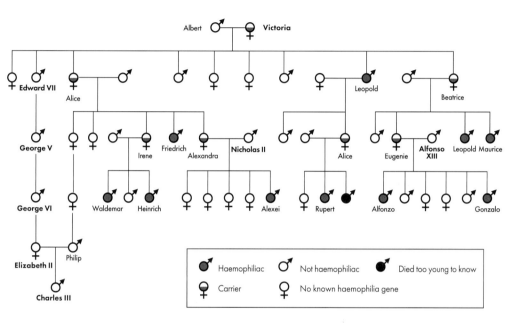

Royal haemophilia

Of Victoria's five daughters, three (at least) inherited the gene. Princess Alice of Hesse passed it on to her son, Prince Friedrich, who died in infancy, and to two daughters, Irene and Alexandra, who passed it on to three haemophiliac grandsons of Alice, including the Tsarevich Alexey of Russia. Irene married her first cousin Henry, a common practice among royals and generally not a good idea because of inbreeding depression. But inbreeding depression was not responsible for the fact that two of their sons, Waldemar and Heinrich, suffered from haemophilia: they got it on their X-chromosome from their mother, and she'd have been equally likely to pass it on, whomever she married, cousin or not (unless the cousin was himself haemophiliac, in which case 50 per cent of her daughters would actually suffer from the disease itself). Another of Victoria's daughters, Princess Beatrice bequeathed the gene to her daughter the Queen of Spain, and on into the Spanish Royal Family, to the resentment, I gather, of the Spanish.

Tracing back the gene tree of the royal haemophilia gene, all lines *coalesce* in Victoria. And indeed, there is a flourishing branch of mathematical genetic theory called Coalescent Theory in which you look back at the history of a genetic variant in a population and trace the most recent common ancestor of that gene – the coalescent gene upon which all lines converge as you look back. Forget about individuals, look through the skin to the genes within, and you can trace two copies of a particular gene back in time until you hit the ancestor in whom they coalesce. That coalescence point is the ancestral individual in which the gene itself divided into two copies, which then went their separate ways in two siblings and eventually two lines of descendants. If you make purifying assumptions like random mating, no natural selection, and everybody has two children, the coalescent tree has an expected form that mathematicians can calculate in theory. In reality, of course, those assumptions are violated, and that's when it becomes interesting. Royal families, for example, typically violate the assumption of random mating. Protocol and political expediency constrain them to marry each other.

Coalescent theory is an important part of modern population genetics, and very relevant to this chapter on the backwards gene's-eye view, but the mathematics is outside my scope here. I will discuss one intriguing example: a particular study of one man's genome – as it happens, my genome, although that isn't why I find it intriguing. It is a remarkable fact that you can make powerful inferences about the demographic history of an entire population using the genome of just a single individual. For a rather odd reason, I was one of the earliest people in Britain to have their entire genome (as opposed to the relatively small sample done by the likes of '23-and-Me') sequenced. I handed the data disc over to my colleague Dr Yan Wong, and he included a clever analysis of it in the book that we co-authored, *The Ancestor's Tale* (2016). It's rather tricky to explain, but I'll do my best.

In every cell of my body swim twenty-three chromosomes inherited intact from my father and twenty-three from my mother. Every (autosomal) paternal gene has an exact opposite number (allele) on the corresponding maternal chromosome, but my father John's chromosomes and my mother Jean's chromosomes float intact and aloof from each other in all my cells. Now, here's where it gets tricky. Take a particular gene on a John chromosome and allow it to look back at its ancestral history. Now take its opposite number ('allele') on the equivalent Jean chromosome, and allow it to look back in the same way. It's the same principle as tracing the royal haemophilia gene back to Victoria. But, in this case, it is not haemophilia that is being traced, we're looking a lot further back, and we have no hope of identifying a named individual like Victoria. We could do it with any pair of alleles, one on a John chromosome and the other on a Jean chromosome. And not just one such pair but (a sample of the) many.

Sooner or later, each gene pair, as they look back, is bound to converge on a particular individual in whom a gene once split to form the ancestor of the John gene and the ancestor of the Jean gene. I really do mean a particular individual ancestor who lived at a particular time and in a particular place. This individual had two children, one of whom was John's ancestor and the other Jean's ancestor. But we're talking about a different ancestral individual – different time and place – for each Jean/John gene pair. For each gene pair, there must have been two siblings, one carrying the ancestral Jean gene and the other the ancestral John gene.

There are many overlapping *people*-tree routes that trace my father and my mother back to different shared ancestors. But for each of my John *genes* there is only one path linking it to the shared ancestor of my corresponding Jean gene. Gene trees are not the same as people trees. Each gene pair coalesces in a particular ancestor, at a particular moment in the past. You can let each pair of my genes look back, and you can find a different coalescence point in each case. You can't literally identify the exact coalescence point for any given gene pair. But what you can do, using the mathematics of coalescent theory, is estimate *when* it occurred. When Dr Wong did this with my genome, he found that a large majority coalesced somewhere around 60,000 years ago, say 50,000 to 70,000.

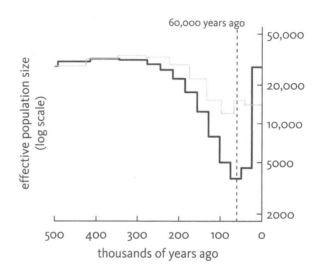

And how should this concordance be interpreted? It means that my ancestors suffered a population bottleneck around that time. Very likely, yours did too. As my John genes and my Jean genes look back at their history, during most of those millennia they see a picture of outbreeding. But somewhere around 60,000 years into the past, the effective population size narrowed to a bottleneck. When the population is smaller, the Jean and John lineages are more likely to find themselves in a shared ancestor, simply by chance. That is why my gene pairs tend to coalesce around that time. Indeed, the coalescence data from my genome, on its own, making use of no other data, can be translated into the above graph of effective population size plotted against time. It is presumably typical for Europeans. The faint grey line shows the equivalent for an individual Nigerian, whose ancestors, it would seem, were not subject to the same bottleneck. I confess to an obscure satisfaction that, of the two co-authors of a book, one was able to use the genome of the other to make a quantitative estimate of prehistoric demography affecting not just one individual but millions.

What else can genes tell us as they look back at their history? Zoologists are accustomed to drawing family trees of animals, and calculating which species are close cousins of other species, and which distant. Among ape species, for example, chimpanzees and bonobos are our closest living relatives, and those two species are exactly equally close to us. They are equally close because they share an ancestor with each other some 3 million years ago, and that ancestor shares an ancestor with us about 6 million years ago (see below).

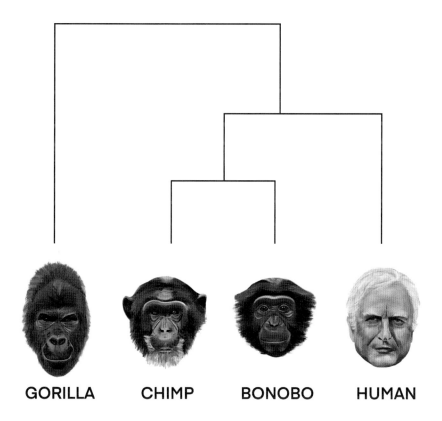

GORILLA   CHIMP   BONOBO   HUMAN

Gorillas are the outgroup, a more distant relative of the rest of us African apes. The ancestor we share with gorillas lived longer ago, perhaps 8 or 9 million years.

On the previous page is the conventional way to draw a family tree, an organism-based family tree. But we can also draw a family tree from the point of view of a gene, looking back at its own history. The organism tree is unequivocal. Chimps and bonobos are close cousins of each other, and we are their closest relatives apart from each other. But while that is indeed a fact from the point of view of the whole organism, it is not necessarily the case when it is genes that look in the rear-view mirror. True, a majority of genes would 'agree' with each other and with the 'people tree' of the traditional zoologist. Nevertheless, it is perfectly possible that, from the point of view of some particular genes, the family tree could look very different. As on the opposite page, perhaps. The majority of our genes agree with the 'people tree'. But when the gorilla genome was published in 2012, it turned out that 'Humans and chimpanzees are genetically closest to each other over most of the genome, but the team found many places where this is not the case. Fifteen per cent of the human genome is closer to the gorilla genome than it is to chimpanzee, and 15 per cent of the chimpanzee genome is closer to the gorilla than human.' I hope you agree that his kind of conclusion is an interesting product of the 'backward gene's-eye view'.

Such an anomaly could occur even within one small family. Two brothers, John and Bill, share the same parents, Enid and Tony, and the same four grandparents: Arthur and Gertrude, the parents of Enid, and Francis and Alice, the parents of Tony. (Sex chromosomes apart) each of the brothers received *exactly* half his genes from each of their shared parents. That's because each is the product of exactly one egg from Enid and one sperm from Tony. And each brother received a quarter of his genes from each of the four shared grandparents, but in this case the figure is only *approximate*. It's not *exactly* a quarter. Through the vagaries of chromosomal crossing-over, the sperm from Tony that conceived John could, by chance, have contained mostly Alice's genes rather than Francis's. The sperm from Tony that con-

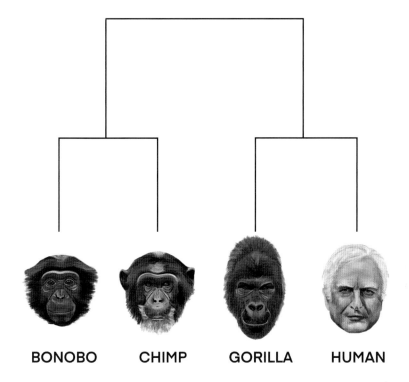

BONOBO   **CHIMP**   **GORILLA**   **HUMAN**

ceived Bill could have contained a preponderance of Francis's genes rather than Alice's. The egg from Enid that gave rise to John could have contained mostly Arthur's genes, while the egg from Enid that gave rise to Bill contained a preponderance of Gertrude's genes. It's even theoretically possible (though vanishingly improbable) that John received all his genes from two of his grandparents, and none from the other two. Thus, the gene's-eye view of closeness of relatedness can differ from the individual's-eye view. The individual's-eye view sees all four grandparents as equal contributors.

And the same is true of all generations prior to the immediate parental generation. Although you are quite probably descended from William the Conqueror, it is also quite likely that you have inherited not a single gene from him. Biologists tend to follow the historic precedent of tracing ancestry at the level of the whole individual organism: every individual has one father and one mother, and so on back. But the John/Bill, gorilla/chimpanzee comparison of the previous paragraphs will prove, I believe, to be the tip of an iceberg. More and more, we shall see pedigrees being drawn up from the genes' point of view as opposed to the individual organism's. An example is the discussion of the prestin gene in Chapter 5. Such a trend is obviously highly congenial to this book, stressing, as it does, the gene's-eye view.

The last topic I want to deal with in this chapter on the backwards gene's-eye view is Selective Sweeps. Among the messages from the past that the genes of a living animal whisper to us, if only we could hear them, many tell of ancient natural selection pressures. That, indeed, is what I mean by the genetic book of the dead, but here I am talking about a particular kind of signal from the past, one that geneticists have learned how to read. Present-day genes send statistical 'signals' of natural selection pressures. A gene pool that has recently undergone strong selection shows a certain characteristic signature. Natural selection leaves its mark. A Darwinian signature. Here's how.

Two genes that sit close to one another on a chromosome tend to travel together through the generations. This is because chromosomal crossing over is relatively unlikely to split them: a simple consequence of their proximity to each other. If one gene is strongly favoured by natural selection it will increase in frequency. Of course, but mark the sequel. Genes whose chromosomal position lies close to a positively selected gene will also increase in frequency: they 'hitch-hike'. This is especially noticeable when the linked genes are neutral – neither good nor bad for survival. When a particular region of a chromosome contains a gene that is under strong selection in its favour, the geneticist notices a diminution in the amount of variation in the population, specifically in the hitch-hiking zone of the affected chromosome. Because of the hitch-hiking, natural selection of one favoured gene 'sweeps' away the variation among nearby neutral genes. This 'selective sweep' then shows up as a 'signature' of selection.

I find the 'backwards' way of looking at ancestral history illuminating. But the most important 'experience' that a gene can 'look back on' is easily overlooked because it hides in plain view. It is the *companionship* of other genes of the species: other genes with which it has had to share a succession of bodies. I am not talking here about genes being linked close to each other on the same chromosome. I am now talking about shared membership of the same gene pool, and hence of many individual bodies. This companionship is the topic of the next chapter.

12

Good Companions, Bad Companions

The previous chapter could be expanded with an indefinite number of examples of the backward gene's-eye view. Genes look back on a series of environments variously characterised by trees, soil, predators, prey, parasites, food plants, water holes, etc. But the external environment is only part of the story. It leaves out the most important kind of 'experience' of a gene. Far more important is the experience of rubbing shoulders with all the other genes in a long succession of bodies: partners through dynasties of mutual collaboration in the subtle arts of building bodies. That is the central point of this chapter.

The genes within any one gene pool are travelling bands of good companions, journeying together, and cooperating with each other down the generations. Genes in other gene pools, gene pools belonging to other species, constitute parallel bands of travelling companions. These bands do not include the genes of other species. That is precisely how biologists like to define a species (although the definition sometimes blurs in practice, especially when new species are being born).

Sexual reproduction validates the very notion of a species, more precisely the notion of a gene pool: a pool of genes like a stirred pool of water. The gene pool is thoroughly stirred in every generation by sexual reproduction, but it doesn't mix with any other such pool – pools belonging to other species. Children resemble their parents but, because the gene pool is stirred, they resemble them only slightly more than they resemble any random member of the species – and much more than they resemble a random member of another species. The gene pool of each species sloshes about in a watertight compartment of its own, isolated from all others.

As I said, that is part of the very definition of a 'species', at least the most widely adopted definition, the one codified by that lofty patriarch among evolutionists, Ernst Mayr (1904–2005):

> Species are groups of actually or potentially interbreeding natural populations, which are reproductively isolated from other such groups.

Fossils, being dead to the possibility of actually interbreeding – beyond breeding at all – force a retreat to Mayr's 'potentially'. When we say that *Homo erectus* was a separate species, distinct from modern *Homo sapiens*, the Mayr definition would be interpreted as meaning, 'If a time machine enabled us to meet *Homo erectus*, we would be incapable of interbreeding with them.' A niggling difficulty arises over 'incapable'. There are species that can be persuaded to interbreed in captivity but would not choose to do so in the wild. Chapter 9's example of the two crickets *Teleogryllus oceanicus* and *commodus* is only one of several. Even if we were capable of interbreeding with *Homo erectus*, say by artificial insemination, would we – or they – choose to do so by the normal, natural means? Never mind, that is a detail that might concern a pernickety taxonomist or philosopher, but we can pass it by.

If, as most anthropologists believe, we descend from *Homo erectus*, there must have been intermediates during the transitional phase: intermediates that would defy classification. Nobody who has thought it through would suggest that suddenly a *sapiens* baby was born to proud *erectus* parents. Every animal ever born throughout evolutionary history would have been classified in the same species as its parents, not only by the interbreeding criterion but by all sensible criteria. That fact – though it troubles some minds – is totally compatible with the fact that *Homo sapiens* is descended from *Homo erectus*, those two species being distinct species incapable – let us presume – of breeding with each other. It's also compatible with the fact that you are descended from a lobe-finned fish, with every intermediate along the way being a member of the same species as its parents and its children.

Moreover, when a species splits into two daughter species in the process known as *speciation*, there is bound to be an interregnum when the two are still capable of interbreeding. The split originates accidentally, imposed perhaps by a geographic barrier such as a mountain range or a river or stretch of sea. It is probable that chimpanzees and bonobos started to go their separate evolutionary ways when two sub-populations found themselves on opposite sides of the Congo river. The two populations were physically prevented from interbreeding – the flow of genes was halted by the flow of water between them. For a while, they could *potentially* interbreed, and maybe occasionally did so when an individual inadvertently crossed the river on a floating log. But the geographically imposed lack of gene flow freed them to evolve in separate directions. Those different directions could have been guided by natural selection, or unguided in a process of random drift. It doesn't matter, the point is that the compatibility between their genes gradually declined until a stage was reached when, even if they should chance to meet, they could no longer interbreed in actuality. The initial geographic barrier doesn't necessarily come about through an environmental change like an earthquake diverting a river. Geography can stay the same while a pregnant female, for instance, gets accidentally washed ashore on a deserted island. Or the other side of a river.

But why, in any case, do the genes of two separated populations tend to become incompatible as companions, thereby preventing interbreeding? One reason is that the two sets of chromosomes need to pair off in the process of meiosis, when gametes are made. If they become sufficiently different, say on opposite sides of a barrier, hybrids, if any, would be unable to make gametes. They might live, but could not reproduce. Another reason – back to the central point of this chapter – is that genes, on either side of the barrier, are naturally selected to cooperate with other genes on the same side, but not the other. After enough time has elapsed in physically enforced separation, two gene pools become so incompatible that interbreeding becomes impossible even if the physical barrier is removed. Chimpanzees and bonobos haven't quite reached that stage. Hybrids can be born in captivity.

There doesn't have to be a distinct barrier, like a river, for geographically based speciation to occur. A mouse in Madrid never meets a mouse in Vladivostok but there could well be continuous local gene flow across the 12,000-kilometre gap between them. Given enough time, their descendants could diverge genetically until they could no longer interbreed even if they should somehow contrive to meet. Speciation would have occurred, the barrier being nothing more than sheer distance rather than an unswimmable river or sea, or an impassable desert or mountain range, and despite continuous gene flow locally across the entire range. We have here the spatial equivalent of the temporal continuum between *Homo erectus* and *Homo sapiens*. In both cases the extremes never meet. Yet in both cases there can be an unbroken chain of intermediates happily breeding all the way across the range: range in space for the example of the mice; range in time for the example of *erectus* and *sapiens*.

Occasionally, the chain of intermediates wraps around in a circle, bites itself in the tail, and we have a so-called 'ring species'. Salamanders of the genus *Ensatina* live all around the four edges of California's Central Valley but don't cross the valley. If you start sampling at the southern end of the valley and work your way up the west side to the north, go eastwards across the north end of the valley, then down the eastern side and back around to your starting point, you notice a fascinating thing. The salamanders all along your route around the edge of the valley can interbreed with their neighbours. Yet they gradually change as you go around, and when you arrive back at your starting point, the 'last' species of the ring cannot interbreed with the 'first'. A ring species is a rare case where you can see laid out in the spatial dimension the kind of evolutionary change that you could see along the time dimension if only you lived long enough.

Such considerations render pointless all heated arguments about whether or not closely related animals, living or fossil, belong to the same species. It is a necessary consequence of evolution that there must be, or must have been, intermediates that you cannot forcibly assign to either species. It would be worrying if it were otherwise. But of course most species in existence are clearly distinct from most other species by any criterion, because of the long time that has elapsed since their ancestors diverged. As for the grey areas where potential interbreeding is even an issue, and where species definition is problematical, this chapter will not treat them further.

Where external environments are concerned, the genes of a mole speak to us of damp, dark, moist tunnels, of earthy smells, of earthworms and beetle larvae crawling between tangled rootlets and filaments of fungal mycelium and mycorrhizae. The genes of a squirrel have a very different ancestral autobiography, a tale of airy greenery, waving boughs, acorns, nuts, and sunlit glades to be crossed between trees. We could weave a similar list for any species. The point of this chapter, on the other hand, is that the genes' external 'experience' of damp, dark soil, or forest canopy, grassy plains, coral reefs, the deep sea, or whatever it might be, is swamped by the more immediate and salient internal experiencing of other genes in the stirred gene pool. This chapter is about the 'good companions' with which the genes have travelled and collaborated, in body after body since earlier times: parting from and re-joining, ever encountering and re-encountering familiar sets of companion genes, collaborating in the difficult arts of building livers and hearts, bones and skin, blood corpuscles and brain cells. The details will be tweaked by 'external' pressures: the best heart, kidney, or intestine for a burrowing vermivore is doubtless not the same as the best heart, kidney, or intestine for a tree-climbing nut-lover. But a centrally important quality of a successful gene will be the ability to collaborate with the other genes of the shared gene pool, be it mole, squirrel, hedgehog, whale, or human gene pool.

Every biochemistry lab has on its wall a huge chart of metabolic pathways, a bewildering spaghetti of chemical formulae joined by arrows. Below is a simplified version in which chemicals are represented by blobs rather than having their formulae spelled out. The lines represent chemical pathways between the blobs. This particular diagram refers to the gut bacterium *Escherichia coli*, but something similar, and just as bewildering, is going on in your cells.

Every one of those hundreds of lines is a chemical reaction performed inside a living cell, and each one is catalysed by an enzyme.

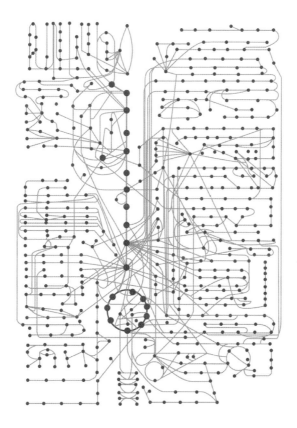

Every enzyme is assembled under the influence of a specific gene (or often two or three genes, because the enzyme molecule may have several 'domains' wrapped around each other, each domain being a protein chain). The genes that make these enzymes must cooperate, must be good companion genes in the sense of this chapter.

All mammals have almost exactly the same set of over 200 named bones, connected in the same order, but differing in size and shape. We saw the principle in the crustaceans of Chapter 6. And the same is true of the metabolic pathways diagrammed above. They are almost the same in all animals but different in detail. And, although they may be engaged in joint enterprises that are similar, the cartels of mutually compatible genes will not be compatible with parallel cartels evolving in other lineages: antelope cartels versus lion cartels, say. Antelopes and lions both need metabolic pathways in all their cells, and both need hearts, kidneys, and lungs, but they'll differ in details appropriate to herbivores versus carnivores. And more obviously so in teeth, intestines, and feet, for reasons we've covered already. If they were somehow to mix in the same body, they wouldn't work well together.

I shall say that two separate gene pools, for instance an impala gene pool and a leopard gene pool, represent two separate 'syndicates' of 'cooperating' genes. Building a body is an embryological enterprise of immense complexity, involving feats of cooperation between all the genes in the active genome. Different kinds of body require different embryological 'skills', perfected over evolutionary time by different suites of mutually compatible genes: compatible with members of their own syndicate but incompatible with other syndicates simultaneously being built in other gene pools. These cooperating cartels are assembled over generations of natural selection. The way it works is that each gene is selected for its compatibility with other genes in the gene pool, and vice versa. So cartels of mutually compatible, cooperating genes build up. It is tempting but misleading to speak of alternative cartels being selected as whole units versus other cartels as whole units. Rather, cartels assemble themselves because each member gene is separately selected for its compatibility with other genes within the cartel, which are themselves being selected at the same time.

Within any one species, genes work together in embryological harmony to produce bodies of the species' own type. Other cartels in other species' gene pools self-assemble, and work together to produce different bodies. There will be carnivore cartels, herbivore cartels, burrowing insectivore cartels, river-fishing cartels, tree-climbing, nut-loving cartels, and so on. My main point in this chapter on 'Good Companions' is that by far the most important environment that a gene has to master is the collection of other genes in its own gene pool, the collection of other genes that it is likely to meet in successive bodies as the generations go by. Yes, the external ecosystem furnished by predators and prey, parasites and hosts, soil and weather, matters to the survival of a gene in its gene pool. But of more pressing moment is the ecosystem provided by the other genes in the gene pool, the other genes with which each gene is called upon to cooperate in the construction and maintenance of a continuing sequence of bodies. It is an easily dispelled paradox that my first book, *The Selfish Gene*, could equally well have been called *The Cooperative Gene*. Indeed, my friend and former student Mark Ridley wrote a fine book with that very title. In his words, which I'd have been pleased to have written myself,

> The cooperation between the genes of a body did not just happen. It required special mechanisms for it to evolve, mechanisms that arrange affairs such that each gene is maximally selfish by being maximally cooperative with the other genes in its body.

As inhabitants of today's technologically advanced world, we are aware of the power of cooperation between huge numbers of specialist experts. SpaceX employs some 10,000 people, cooperating in the joint enterprise of launching massive rockets into space and – even more difficult – bringing them back and gently landing them in a fit state to be re-used. Many different specialists are united in intimate cooperation: engineers, mathematicians, designers, welders, riveters, fitters, turners, computer programmers, crane operators, quality control checkers, 3-D printer operators, software coders, inventory control officers, accountants, lawyers, office workers, personal assistants, middle managers, and many others. Most of the experts in one field have little understanding of what experts in other parts of the enterprise do, or how to do it. Yet the feats that we humans can achieve when thousands of us deploy our complementary skills, in well-oiled collaboration but in ignorance of each other's role, are staggering.

The human genome project, the James Webb Telescope, the building of a skyscraper or a preposterously oversized cruise ship, these are stunning achievements of cooperation. The Large Hadron Collider at CERN brings together some 10,000 physicists and engineers from more than 100 countries, speaking dozens of languages, working smoothly together to pool their diverse expertise. Yet these huge accomplishments of mass cooperation are more than matched by the nine-month collaborative enterprise of building each one of us in our mother's womb: a feat of cooperation among billions of cells, belonging to hundreds of cell types (different 'professions'), orchestrated by about 30,000 intimately cooperating genes, exceeding the personnel count we find in a large human enterprise such as SpaceX. Cooperation is key, in both building a body and building a rocket.

The genes that build a body must cooperate with all the other companions that the sexual lottery throws at them as the generations go by. They must cooperate not only with the present set of companions, those in today's body. In the next generation, they'll have to cooperate with a different sample of companions drawn from the shared gene pool. They must be ready to cooperate with all the alternative genes that march with them down the generations within this gene pool – but no other gene pool. This is because Darwinian success, for a gene, means long-term success, travelling through time over many generations, in many successive bodies. They must be good travelling companions of all the genes in the stirred gene pool of the species.

The 1957 film of JB Priestley's novel *The Good Companions* had an accompanying song with a not uncatchy tune, of which the refrain was,

> *Be a good companion,*
> *Really good companion,*
> *And you'll have good companions too.*

It is a song whose evoked mutualism suits the travelling troupe of genes, which constitutes the active gene pool of a species such as ours. Sexual recombination of genes gives meaning to the very existence of the 'species' as an entity worth distinguishing with a name at all. Without it, as is the case with bacteria, there is no distinct 'species', no clear way to divide the population with confidence into discrete nameable groups. It is sexual reproduction that confers identity on the species. Some bacterial types are not far from being a big smear, grading into each other as they promiscuously share genes. The attempt to assign discrete species names to such bacteria is a losing battle in a way that doesn't apply to animals like us, where sexual exchange is limited to sexual encounters between a male and a female of the same species – and no other species by definition. As already stated, where fossils are concerned we have to guess, based on their anatomical similarity, whether they would have been able to interbreed when they were alive. This involves subjective judgement, which is why naming fossils such as *Homo rhodesiensis* and *Homo heidelbergensis* is a matter of aggravated controversy between 'lumpers' and 'splitters'. But notwithstanding naming disagreements, which can even become acrimonious, we remain confident that the gene pool surrounding every one of those fossils was a troupe of travelling companions isolated from other gene pools – even though imperfectly isolated during episodes of speciation. Bacteria largely deny us that confidence. So-called 'species' of bacteria are not clearly delimited.

Every working gene, 'expert' in rendering up its own contribution to the collaborative building of an embryo, is confined to its own gene pool. Repeated cooperation among successive samples drawn from the same troupe of travelling companions has selected genes largely incapable of working beneficially with members of other troupes. Not entirely, as we see from headlined examples like jellyfish genes transplanted into cats and making them glow in the dark. Genes are normally not put to that kind of test. Mules and hinnies, ligers and tigons, are almost always sterile. Their sets of travelling companions are still compatible enough to collaborate in building strong bodies.

But their compatibility breaks down when it comes to chromosomal pairing-off in meiosis, the process of cell division that makes gametes. Mules can pull a cart, but they can't make fertile sperms or eggs.

Nature doesn't transplant antelope genes into leopards. If it did, a few might work normally. There are broad similarities between the embryologies of all mammals, and all mammals doubtless share genes for making most layers of the mammalian palimpsest. But that doesn't undermine this chapter. Those genes concerned with what makes a leopard a predator, and an antelope its herbivorous prey, would not work harmoniously together. In childishly crude terms, leopard teeth wouldn't sit well with antelope guts and antelope feeding habits. Or vice versa. In the language of this chapter, companions that travel well together in one gene pool would not be *good* companions in the other. The collaboration would fail.

The principle is illustrated by an old experiment of EB Ford, the eccentrically fastidious aesthete from whom I learned my undergraduate genetics. Most practical geneticists work on lab animals or plants, breeding fruit flies or mice in the laboratory. But Ford walked a minority path among geneticists. He and his collaborators monitored evolutionary change in gene pools, in the wild. A lifelong authority on butterflies and moths, he went out into the woods and fields, heaths and marshes of Britain, waving his butterfly net and sampling wild populations. He inspired others to do the same kind of thing with wild fruit flies, wild snails and flowers, as well as other species of butterflies and moths. He founded a whole discipline called *Ecological Genetics* and wrote the book of that title. The piece of work that I want to talk about here was a field study of wild populations of lesser yellow underwing moths, in Scotland and some of the Scottish islands. Ford knew it as *Triphaena comes*, but it is now called *Noctua comes*, following the strict precedence rules of zoological nomenclature.

The species is polymorphic, meaning there are at least two genetically distinct types coexisting in significant proportions in the wild. Not in England, however, nor in much of mainland Scotland, where all the lesser yellow underwings look like the pale upper one in the

Dark and light morphs of lesser yellow underwing

picture. But in some of the Scottish islands there exists, in significant numbers, a second morph, of darker colour, called *curtisii*, evidently named after the entomologist and artist John Curtis (1791–1862).

I thought it fitting to use Curtis's own painting of the *curtisii* morph and the cowslip, and I asked Jana Lenzová to paint in the light morph to complete the picture.

The difference between the two morphs is controlled by a single gene, which we can call the *curtisii* gene. *Curtisii* is nearly dominant. This means that if an individual has either one *curtisii* gene ('heterozygous') or two *curtisii* genes ('homozygous for *curtisii*'), it will be dark. If dominance were complete, heterozygous individuals with one *curtisii* gene would look exactly the same as homozygotes with two. *Curtisii* being only nearly dominant, the heterozygotes are almost the same as the *curtisii* homozygotes but slightly lighter. Heterozygotes are always darker than individuals homozygous for the standard *comes* gene, which is therefore called recessive.

Like his mentor Ronald Fisher, whom we've already met, Ford liked to speak of 'modifiers', genes whose effect is to modify the effects of other genes. According to Fisher's theory of dominance, to which Ford subscribed, when a gene first springs into existence by mutation, it is typically neither dominant nor recessive. Natural selection subsequently drives it towards dominance or recessiveness via the gradual accumulation, through the generations, of modifiers. Dominance is not a property of a gene itself, but a property of its interactions with its companion modifiers.

Modifiers don't change the major gene itself. What they change is how it expresses itself, in this case its degree of dominance. The language of this chapter would say that a major gene such as *curtisii* has modifiers among its 'good companions', which affect its dominance, meaning its tendency to express itself when heterozygous. For reasons we needn't go into, natural selection favoured a significant proportion of dark *curtisii* morphs on certain Scottish islands. And one way this favour showed itself, according to the theory of Fisher and Ford, was by selection in favour of modifiers that increased its dominance.

Barra is an island in the Outer Hebrides, west of Scotland. Orkney, north of Scotland, is an archipelago 340 kilometres from Barra as the crow flies, and too far for the moth to fly. Ford collected and studied moths from both these locations. Both have mixed populations of Lesser Yellow Underwings, the normal pale form living alongside significant numbers of dark *curtisii* morphs. Breeding experiments, with both Barra and Orkney moths, separately confirmed the dominance of *curtisii* within both islands. However, when Ford took moths from Barra and crossed them with moths from Orkney, he got a remarkable result. The dominance broke down. It disappeared. No longer did Ford see tidy Mendelian segregation of dark versus light forms. Instead there was a messy spectrum of intermediates. Dominance had disappeared.

What had evidently happened was this. Dominance on Barra had evolved by an accumulation of mutually compatible modifiers, good Barra companions. Dominance on Orkney had independently and convergently evolved by a different consortium of modifier genes, good Orkney companions. When Ford bred across islands, the two sets of modifiers couldn't work together. It was as though they spoke different languages. To work properly, each modifier needed its normal set of good companions, the set that had been built up over generations of selection on the different islands. That's what being good companions is all about, and Ford's experiment dramatically demonstrates a principle that I believe to be general. The 'major' gene, *curtisii*, is the same on both Barra and Orkney. However, for all that a gene itself is the same, its dominance can be built up in more than one way by different consortia of modifiers. This seems to have been the case with *curtisii* on different islands.

There's a potential fallacy lurking here. It's easy to presume that the Barra good companions lie close to each other on a chromosome and therefore segregate as a unit. And likewise, the Orkney consortium of good companions. That kind of thing can happen, and Ford and his colleagues discovered it in other species. Natural selection can favour inversions and translocations of bits of chromosome that bring good companions closer to each other.

Sometimes they end up so close that they are called a 'supergene', so close that they are rarely separated by crossing over. This is an advantage, and the translocations and inversions that contribute to the building of a supergene are favoured by natural selection. But if Ford's modifiers had been clustered together as a supergene in the case of his yellow underwings, he wouldn't have got the results that he did.

Supergenes can be demonstrated in the lab by breeding large numbers of individuals for many generations until suddenly, by a freak of chromosomal crossing-over, the supergene is split. But the supergene phenomenon is not necessary for good companionship, and there's no reason to suppose it applies in this case of the lesser yellow underwing. The suites of cooperating modifiers could lie on different chromosomes all over the genome. Separately, in their respective island gene pools, they were assembled by natural selection as good team workers in each other's presence. In this case, they work well together to increase the dominance of the *curtisii* gene. But the principle is more general than that. We don't have to subscribe to the Fisher/Ford theory of dominance in particular.

Natural selection favours genes that work together in their own gene pool, the gene pool of their species. Genes that go with being a carnivore (say, genes for carnivorous teeth) are naturally selected in the presence in the same gene pool of other 'carnivorous genes' (say genes for short carnivorous intestines whose cells secrete meat-digesting enzymes). At the same time, on the herbivore side, genes for flat, plant-milling teeth flourish in the presence of genes for long, complicated guts that provide havens for plant-digesting micro-organisms. Once again, the alternative suites of genes may be distributed all over the genome. There's no need to assume that they cluster together on any particular chromosome.

Unfortunately, good companionship sometimes breaks down. It is even subject to sabotage. We've already met ways in which the genes within a body can be in conflict with one another. The uneasy pandemonium of genes within the genome, sometimes cooperating, sometimes disputing, is captured in Egbert Leigh's 'Parliament of

Genes'. Each acts 'in its own self-interest, but if its acts hurt the others, they will combine together to suppress it.'

Cell division within the body is vulnerable to occasional 'somatic' mutation. Of course it is. How could it not be? We are familiar with the idea that random copying errors, mutations, produce the raw material for natural selection between individuals. Those 'germline' mutations occur in the formation of sperms and eggs, and they are then inherited by an individual's children. These are the mutations that play an important role in evolution. But most acts of cell division occur *within* the body – somatic as opposed to germline mutation – and they too are subject to mutation. Indeed, the mutation rate per mitotic division is higher than for meiotic division. We should be thankful our immune system is so good at spotting the danger early. Most somatic mutations, like most germline mutations, are not beneficial to the organism. Sometimes they are beneficial to themselves but bad for the organism, in which case they may engender malignant tumours – cancers. Subsequent natural selection within the tumour can generate a progression through increasingly ominous 'stages' of cancer. I shall return to this.

We can think of the (somatic) cells in a developing embryo as having a family history within the body, springing from their grand ancestor, the single fertilised egg cell of a few months or weeks previous. At any stage in this history of descent, starting with the embryo and on throughout the rest of life, somatic mutation can occur. Vertebrate development is the product of countless cell divisions, so embryologists have found it convenient to trace cell lineages in a simpler organism. The tiny roundworm *Caenorhabditis elegans* has only 959 cells. It was the genius of the great molecular biologist Sydney Brenner to pick this animal out as the ideal subject for a genre of research that has since spread to dozens of labs throughout the world. Its embryo at one of its developmental stages has precisely 558 cells. Every one of those 558 cells has its own 'ancestral' sequence within the developing embryo. The pedigree of each of those 558 cells within the embryo has been painstakingly worked out (next page). Necessarily, it's impossible to print the details legibly on one page of a book, but you can expand it here (https://www.wormatlas.org/celllineages.html) and get an idea of the diverging pedigree of cells in the embryo, consisting of 'families' and 'sub-families'. If you could read the labels by the side of families of cells, you'd see things like 'intestine', 'body muscle', 'ring ganglion'. We shall have need to return to that idea of families of cells procreating in the embryo.

Now, if that's what the cellular pedigree looks like for a mere 558 roundworm cells, just think what it must look like for our 30 to 40 trillion cells. Similar labels – muscle, intestine, nervous system, etc. – could be affixed to cells in a human embryo (opposite). This is true even though the pedigrees are not determined so rigidly in a vertebrate embryo, and we can't enumerate a finite tally of named cells. It's important to stress that these differ-ent families of cells within the developing embryo are, until something goes wrong, genetically identical. If they weren't, they might not cooperate. When something goes wrong and they're no longer genetically identical, well that's when there's a risk of their becoming bad compan-ions. And then there's a risk of their evolving, by natural selection within the body, to become very bad compan-ions indeed: cancers.

As you can see on the diagram on the facing page, after some early cell generations within the embryo, the pedigree of our cells splits into three major families: the ectoderm, the mesoderm, and the endoderm. The ecto-derm family of cells is destined to give rise, further down the line, to skin, hair, nails, and those hugely magnified nails that we know as hooves. Ectodermal derivatives also contribute the various parts of the nervous system. The endoderm family of cells branches to give rise to sub-families that eventually make the stomach and intes-tines; and other sub-families that make the liver, lungs, and glands such as the pancreas. The mesoderm dynasty of cells spawns numerous sub-families, which branch

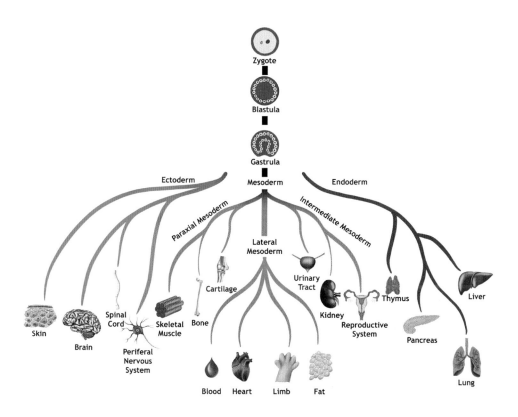

again and again to produce muscle, kidney, bone, heart, fat, and the reproductive organs, although not the germline, which is early hived off and sequestered for its privileged destiny, on down the generations.

Somatic mutants apart, every one of the cells in the expanding pedigree has the same genome, but different genes are switched on in different tissues. That is to say they are *epigenetically* different while being *genetically* the same (see the relevant endnote if popular hype has confused you as to the true meaning of 'epigenetics'). Liver cells have the same genes as muscle cells, but once they pass beyond a certain stage in embryonic development, only liver-specific genes are active there. And the liver 'family' of cells in the pedigree goes on dividing until the liver is complete. They then stop dividing. The same applies to all the 'families', which each have their own stopping time. Cells must 'know' when to stop dividing. And that is where trouble can step in.

With an important reservation, the number of cell generations before the arresting of cell division varies from tissue to tissue and is typically between forty and sixty. That may seem surprisingly few. But remember the power of exponential growth. Fifty liver cell generations, if each one was a division into two (fortunately it isn't) would yield a liver the size of a large elephant. Different cell lines stop dividing after different limits, producing end organs of different sizes. You can see how important it is for each cell line to know when to stop dividing.

Cactus with somatic mutation

Every one of the 30 trillion cells in a body was made by a cell division. And every one of those cell divisions is vulnerable to somatic mutation. Now we come to that 'important reservation', the one relevant to the topic of bad companions. The cells in a lineage are genetically identical only if no somatic mutation intervenes during the lineage's successive generations. Most somatic mutations are harmless. But what if a somatic mutation arises in a cell such that it changes its behaviour and refuses to stop dividing? Its lineage in the 'family tree' doesn't come to a disciplined halt, but goes on reproducing out of control. The daughter cells of the mutant cell inherit the same rogue mutation, so they too divide. And their daughter cells inherit the rogue gene, so ... This is the kind of thing that produces weird growths such as adorns the cactus opposite.

Let's follow the subsequent history of a rogue cell's descendants, for example in a human. Reproducing for an indefinite number of generations without discipline, these cells will now be subject to a form of natural selection. Why say 'a form of'? It is natural selection, plain and simple. The rogue cells will be subject to natural selection, every bit as Darwinian as the natural selection that chooses the fastest pumas or pronghorns, the prettiest peacocks or petunias, the most fecund codfish or dandelions. Rogue somatic mutant cells can evolve, by natural selection within the body, into cancers that spread menacingly ('metastasis') to other parts of the body. Now natural selection of cells within the tumour will favour those that become better cancers. What does 'better' mean, for a cancer? They become expert, for example, at usurping a large blood supply to nurture themselves. The whole subject, fascinating, disturbing, and not at all surprising to a Darwinist, is expounded in books such as Athena Aktipis's *The Cheating Cell*, and *The Society of Genes* by Itai Yanai and Martin Lercher.

Since cancers evolve by natural selection (within the body), we should treat their evolutionary adaptations in just the same way we might treat the adaptations of pronghorn or codfish, except that the ecological environment is the interior of a (say) human body instead of the sea or an open prairie. This chapter's discussion of Good Companions has prepared us for the idea of an ecology of genes within the body, to parallel the more conventional idea of an external ecology. And that internal ecology is also the setting where *bad* companions can thrive. An important difference is that natural evolution in the open sea or prairie goes on into the indefinite future. The evolution of a cancer tumour ends abruptly with the death of the patient, whether that death is caused by the cancer or something else. The cancer evolves to become better and better at (as an inadvertent by-product) killing itself. This, too, should not surprise. Natural selection, as I've said over and over, has no foresight. A tumour cannot foresee that increased malignancy will eventually kill the tumour itself. Natural selection is the *blind* watchmaker. Despite ending with the death of the organism, the number of generations of cell division in a tumour is large enough to accommodate constructive evolutionary change. Constructive from the point of view of the cancer. Destructive for the patient. Athena Aktipis's book artfully treats the evolution of cancer cells in the body in just the kind of way we might treat the evolution of buffalos or scorpions in the Serengeti.

Cancer cells, then, or rather the mutant genes that turn cells cancerous, are one kind of 'bad companion'. Another type is the so-called *segregation distorter*. Sperms and eggs – gametes – are 'haploid' cells you'll remember, having only one copy of each gene, instead of two like normal body cells. The special kind of cell division called meiosis makes haploid gametes (having only one set of chromosomes) out of diploid cells, which have two sets of chromosomes, one set from the individual's mother and another set from the father. It is only when gametes are made by meiosis that the two sets meet each other in the same chromosome. Meiosis performs an elaborate shuffle, cutting and pasting exchanged portions of paternal and maternal chromosomes into a new set of mixed-up chromosomes. Every gamete is unique, having different assortments of paternal and maternal genes in each of its (twenty-three in humans) chromosomes. The result of the shuffle is that each gene from the diploid set of (forty-six in humans) chromosomes has a 50 per cent chance on average of getting into each gamete.

The 'phenotypic effect' of a gene commonly shows itself somewhere in the body – it might affect tail length or brain size or antler sharpness. But what if a gene were to arise that exerted its phenotypic effect on the process of gamete production itself? And what if that effect was a bias in gamete production such that the gene itself had a greater than 50 per cent chance of ending up in each gamete? Such cheating genes exist – '*segregation distorters*'. Instead of the meiotic shuffle resulting in a fair deal to each gamete as it normally does, the deal is biased in favour of the segregation distorter. The distorter gene has a greater than even chance of ending up in a gamete.

You can see that if a rogue segregation distorter were to arise it would tend, other things being equal, to spread rapidly through the population. The process is called meiotic drive. The rogue gene would spread, not because of any advantage to the individual's survival or reproductive success, not because of benefit of any kind in the conventional sense, but simply because of its 'unfair' propensity to get itself into gametes. We could see meiotic drive as a kind of population-level cancer. A special case of a segregation distorter is the 'driving Y-chromosome', that is, a gene on a Y-chromosome whose effect on males is to bias them towards producing Y sperms and therefore male offspring. If a driving Y arises in a population, it tends towards driving it extinct for lack of females: population-level cancer indeed. Bill Hamilton even suggested that we could control the yellow fever mosquito by deliberately introducing a driving male into the population. Theoretically, the population should drastically shrink through lack of females.

Other ways have been suggested to control pests by 'driving genes'. I've already mentioned in Chapter 8 the crass irresponsibility of the 11th Duke of Bedford in introducing grey squirrels, native to America, into Britain. He not only released them in his own estate, Woburn Park, but made presents of grey squirrels to other landowners up and down the country. I suppose it seemed like a fun idea at the time, but the consequence is the wiping-out of our native red squirrel population. Researchers are now examining the feasibility of releasing a driving gene into the grey squirrel gene pool. This would not be carried on the Y-chromosome but would produce a dearth of females in a slightly different way. The authors of the idea are mindful of the need to be careful. We want to drive the grey squirrel extinct in Britain but not in America where it belongs and where it would have stayed but for the Duke of Bedford.

Bad companions, at least in the form of cancers, force themselves upon our forebodings. But for our purposes in this book, it is the gene's role as good companion that we must thrust to prominence. It remains for the last chapter to pin down exactly what makes them cooperate. Fundamentally, it is, I maintain, the fact that they share an exit route from each body into the next generation.

Good companions dressed for field work: RA Fisher and EB Ford. See endnote for my suspicion that this is a historic photogaph.

13

Shared Exit to the Future

Purveyors of scientific wonder like to surprise us with the prodigious – disturbing to some – numbers of bacteria inside our bodies. We're accustomed to fearing them but most of them are, in the words of Jake Robinson's title, *Invisible Friends*. Mostly in the gut, estimates vary from 39 trillion to 100 trillion, the same order of magnitude as the number of our 'own' cells, where 40 trillion is a round-number estimate. Between a half and three-quarters of the cells in your body are not your 'own'. But that doesn't take account of the mitochondria. These miniature metabolic engine-rooms swarm inside our cells and the cells of all eucaryotes (that is, all living creatures except bacteria and archaea). It is now established beyond doubt that mitochondria originated from free-living bacteria. They reproduce by cell division like bacteria, and each has its own genes in a ring-shaped chromosome, again like bacteria. In fact, let us not mince words, they *are* bacteria: symbiotic bacteria that have taken up residence in the hospitable interior of animal and plant cells. We even know, from DNA-sequence evidence, which of today's bacteria are their closest cousins. The number of mitochondria in your body is many trillions.

The bacteria that became mitochondria brought with them much essential biochemical expertise, the research and development of which was presumably accomplished long before they became incorporated as proto mitochondria. Their main role in our cells is the combustion of carbon-based fuel to release needed energy. Not the violent high-speed combustion of fire, of course, but a slow, orderly, trickle-down oxidation. Not only are you a swarm of bacteria, you couldn't move a muscle, see a sunset, fall in love, whistle a tune, despise a demagogue, score a goal, or invent a clever idea without the unceasing activation of their chemical knowhow, expert tricks cobbled together by natural selection choosing between rival bacteria in a lost pre-Precambrian sea.

The interiors of plant cells swarm with green *chloroplasts*, which also are descended from bacteria (a different group, the so-called cyanobacteria). Like mitochondria, chloroplasts *are* bacteria in every sensible meaning of the word. Again like mitochondria, they brought with them a formidable dowry of biochemical wizardry, in this case photosynthesis. Virtually all life on Earth is ultimately powered by energy radiated from the gigantic nuclear fusion reactor that is the sun. It is captured by photosynthesis in chloroplast-equipped solar panels such as leaves, and is subsequently released in the chemical factories that are mitochondria, in all of us. Solar photons that fall on the sea are captured not by leaves but by single-celled green organisms. Whether on land or at sea, solar energy is the base of all food chains. I think the only exceptions are those strange communities whose ultimate source of energy is hot springs, undersea 'smokers' and such conduits of heat from the Earth's interior.

Our mitochondria couldn't do without us, just as we wouldn't survive two instants without them. We are joined deep in mutual amity. Our genes and their genes are good companions that have travelled in lockstep over 2 billion years, each naturally selected to survive in an environment furnished by the other. Most of the genes that originated in their bacterial forebears have long since either migrated into our own chromosomes or been laid off as redundant. But why are mitochondria, and some other bacteria, so benign towards us, while other bacteria give us cholera, tetanus, tuberculosis, and the Black Death? My Darwinian answer is as follows. It is an example of the take-home message of the whole chapter. *Mitochondrial genes and 'own' genes share the same exit route to the future.* That is literally true if we are female, or if we for the moment overlook the fact that mitochondria in males have no future. The key to companionable benevolence, I shall show, or its reverse, is the route by which a gene travels from a present body into a body of the next generation.

Mitochondria and chloroplasts may be the earliest examples of bacteria being coopted into animals, but they are not the only ones. Here's a much more recent re-enactment of those ancient incorporations, and it is highly congenial to the thesis of the gene's-eye view. The embryonic development of vertebrate eyes requires a protein called IRBP, which facilitates the separation of retinal cells from one another and helps them to see better. In a large survey of more than 900 species, IRBPs were found in every vertebrate examined, plus Amphioxus, a small, primitive creature related to vertebrates, although it lacks a backbone. But of the 685 invertebrate species, the only one with a molecule resembling IRBP was an amphipod crustacean, *Hyalella*. Among plants, a single species, *Ricinus communis*, the castor oil plant, has something like an IRBP. And there's a little cluster of fungi too. Molecules resembling IRBPs are ubiquitous among bacteria.

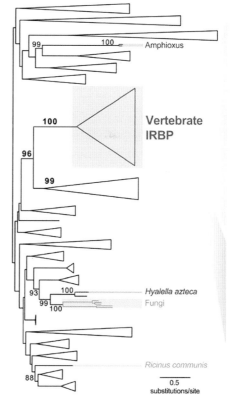

A   **Bacterial S41 peptidases**

99   100   Amphioxus

100   Vertebrate IRBP

96

99

93   100   *Hyalella azteca*
99   100   Fungi

Ricinus communis

0.5
substitutions/site

88

A family tree of IRBP-like molecules shows a richly branched pedigree among bacteria, paralleling that of the vertebrates in which they live, both pedigrees springing from a single point. The isolated pop-ups (crustacean, fungi, and plant) also spring from within the bacterial tree, but widely separated parts of it. This is good evidence of horizontal gene transfer from various bacteria into the eucaryote genome. The evidence strongly suggests that vertebrate IRBPs are 'monophyletic', all descended from a single ancestor, which means a single jump from a bacterium right at the base of vertebrate evolution. Ever since that event, the genes concerned have been passed vertically down the generations. This is like the bacteria that became mitochondria, although mitochondrial ancestors were whole bacteria, not single genes.

I want to give a general name to bacteria that are transmitted from host to host in host gametes: *verticobacter*, because they pass vertically down the generations. The ancestors of mitochondria and of chloroplasts are prime examples of verticobacters. Verticobacters can infect another organism only by riding inside its gametes into its children. By contrast, a typical 'horizontobacter' might pass by any route from host to host. If it lives in the lungs, for instance, we may suppose its method of infection is via droplets coughed or sneezed into the air and breathed in by its next victim. A horizontobacter doesn't 'care' whether its victim reproduces. It only 'wants' its victim to cough (or sneeze, or make bodily contact by hands, lips, or genitals), and it works to that end – 'works' in the sense that its genes have extended phenotypic effects on the host's body and behaviour, driving the host to infect another host. A verticobacter, by contrast, 'cares' very much that its 'victim' shall successfully reproduce, and 'wants' it to survive to reproduce. Indeed, 'victim' is scarcely the appropriate word, which is why I protected it behind quotation marks. This is, of course, because a verticobacter's 'hope' of future transmission lies in the offspring of the host, exactly coinciding with the 'hopes' of the host itself. Therefore, if a verticobacter's genes have extended phenotypic effects on the host, they will tend to agree with the phenotypic effects of the host's own genes. In theory a verticobacter's genes should 'want' exactly the same thing as the host's genes in every detail.

The pertussis (whooping cough) bacterium is a good example of a horizontobacter. It makes its victims cough, and it passes through the air to its next victim, in droplets emitted by the cough. Cholera is another horizontobacter. It exits the body via diarrhoea into the water supply, whence it 'hopes' to be imbibed by somebody else, drinking contaminated water. It doesn't 'care' if its victims die, and it has no 'interest' in their reproductive success.

The notion of a parasite's 'wanting' its victim to do something needs explaining, and this again is where the extended phenotype comes in, as promised at the end of Chapter 8. The parasitology literature is filled with macabre stories of parasites manipulating host behaviour, usually changing the behaviour of an intermediate host to enable transmission to the next stage in the parasite's complicated life cycle. Many of these stories concern worms rather than bacteria, but they convey the principle I am seeking to get across. 'Horsehair worms' or 'gordian worms', belonging to the phylum Nematomorpha, live in water when adult, but the larvae are parasitic, usually on insects. The insect hosts being terrestrial, the gordian larva needs somehow to get into water so it can complete its life cycle as an adult worm. Infected crickets are induced to jump, suicidally, into water. An infected bee will dive into a pond. Immediately the gordian worm bursts out and swims away, the crippled bee being left to die. This is presumably a real Darwinian adaptation on the part of the worm, which means that there has been natural selection of worm genes whose ('extended') phenotypic effect is a change in insect behaviour.

Here's another example, this time involving a protozoan parasite, *Toxoplasma gondii*. The definitive host is a cat, and the intermediate host is a rodent such as a rat. The rat is infected via cat faeces. *Toxoplasma* then needs the infected rat to be eaten by a cat, to complete its life cycle. It insinuates itself into the rat's brain and manipulates the rat's behaviour in various ways to that end. Infected rats lose their fear of cats, specifically their aversion to the smell of cat urine. Indeed, they become positively attracted to cats, though apparently not to non-predatory animals, or predators that don't attack rats. There is some evidence that they lose fear in general, owing to increased production of the hormone testosterone. Whatever the details, it's reasonable to guess that the change in rat behaviour is a Darwinian adaptation on the part of the parasite. And therefore an extended phenotype of *Toxoplasma* genes. Natural selection favoured *Toxoplasma* genes whose extended phenotypic effect was a change in rat behaviour.

The infected snail's bulging eyes are a tempting target for birds

Leucochloridium is a fluke (flatworm), parasitic on birds. Its intermediate host is a snail, and it needs to transfer itself from snail to bird. The snails that it infects are largely nocturnal, while the birds who are the next host feed by day. The worm manipulates the behaviour of the snail to make it go out by day. But that is only the beginning of the snail's troubles. One of the life-history stages of the worm invades the eye stalk of the snail, which swells grotesquely, and seems to pulsate vividly along its length.

This is said to make the eye stalk look like a little crawling caterpillar. Be that as it may, it certainly renders the eye stalks conspicuous, and birds readily peck them off. Infected snails also move around more actively than unparasitised ones. The snail is not killed but only blinded. It is able to regenerate its eye stalks to pulsate another day and perhaps be again plucked off. The fluke, for good measure, castrates its snail victim. And that's an interesting story in its own right. 'Parasitic castration' is common enough to be a named thing.

It is practised by a wide variety of parasites from around the animal kingdom, including protozoa, flatworms, insects, and various crustaceans. Including *Sacculina*, the parasitic barnacle that I introduced in Chapter 6 and promised to return to.

Sacculina is perhaps the most extreme example of the 'degenerative' evolution typical of parasites. Darwin, in his monographs on barnacles, which distracted him for eight of the twenty years when he might have published on evolution, misdiagnosed the affinities of *Sacculina*. And who can blame him? Just take a look at it. The externally visible part of *Sacculina* is a soft bag clinging to the underside of a crab. Most of the 'barnacle' consists of a branching root system permeating the inside of the unfortunate crab's body. Eventually, it fills the body so completely that if you could sweep away the crab and leave only the *Sacculina*, this is what you might see.

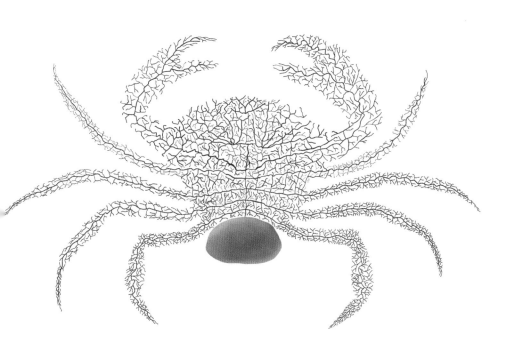

This is not a crab

How do we know that this system of branching rootlets, this sprawling entity that looks like a plant or fungus, is really a barnacle? How do we even know it's a crustacean? The various larval stages of the life cycle give it away. The nauplius larva is followed by the cyprid larva, and both are unmistakeably crustacean. As if final clinching were needed, *Sacculina*'s genome has now been sequenced. 'It is written', as the Muslims say: 'Crustacean'.

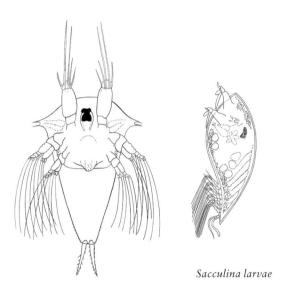

Sacculina larvae

The first organs that *Sacculina* attacks are the crab's reproductive organs. This is the 'parasitic castration' that I mentioned above. Barnacles themselves are sometimes castrated by parasitic crustaceans; marine isopods related to woodlice. So, what is the point of parasitic castration? Why would a parasite head straight for the gonads of its host, before eating other organs?

As with all animals, the host's ancestors have been naturally selected to fine-tune a delicate balance between the need to reproduce (now) and the need to survive (to reproduce later). A parasite such as *Sacculina*, however, has no interest in assisting its host to reproduce. This is because its genes don't share the host genes' exit route to the future. *Sacculina* genes 'want' to shift the host's 'balance' towards surviving, to carry on feeding the parasite. Like a docile, castrated ox being fattened up, the crab is forced by the parasite to renounce reproduction and become a maintained source of food.

The situation reverses in those cases where parasites – 'vertico-parasites' – pass to the next host generation in the gametes of the host. Verticoparasites infect only the offspring of their individual hosts rather than potential hosts at large. The genes of a verticoparasite share the 'exit route' of the host genes, so their extended phenotypic effects will agree with the host genes' phenotypic effects. Exercise our usual cautious licence to personify, and consider the 'preferred options' of a verticoparasite such as a verticobacter. It travels inside the eggs of a host directly into the host's child. Here, the interests of parasite and host coincide, and their genes 'agree' about the optimal host anatomy and behaviour. Both 'want' the host to reproduce, and survive to reproduce. Once again, if the genes of vertically transmit-ted parasites have extended phenotypic effects on their hosts, those effects should coincide, in perfect agreement and in every detail, with the phenotypic effects of the host animal's 'own' genes.

Mitochondria are an extreme example of a verticoparasite. Long transmitted vertically down the generations inside host eggs, they became so amicably cooperative that their parasitic origins are hard to spot, and were long overlooked. A horizontoparasite such as *Sacculina* has opposite 'preferences'. It has no 'interest' in its host's successful reproduction. Whether or not a horizontoparasite 'cares' about its host's survival depends on whether it can benefit from it, presumably, as in the case of *Sacculina*, by feeding on the living host. If, by castration, it can shift the balance of the host's internal economy away from reproduction and towards survival, so much the better.

The tapeworm *Spirometra mansanoides* doesn't castrate its mouse victims but it achieves a similar result. It secretes a growth hormone, which makes them grow fatter than normal mice. And fatter than the optimum achieved by natural selection of mouse genes seeking a balance between growth and reproduction. *Tribolium* beetles normally develop through a succession of six larval moults, increasing in size, before they eventually change into an adult. A protozoan parasite, *Nosema whitei*, when it infects *Tribolium* larvae, suppresses the change to adult. Instead, the larva continues to grow through as many as six extra larval moults, ending up as a giant grub, weighing more than twice as much as the maximum weight of an uninfected larva. Natural selection has favoured *Nosema* genes whose extended phenotypic effect was a dramatic doubling in *Tribolium* fatstock weight, achieved at the expense of beetle reproduction.

A small tapeworm, *Anomotaenia brevis*, needs to get into its definitive host, a woodpecker. It does so via an intermediate host, an ant of the species *Temnothorax nylanderi*, which has the habit of collecting woodpecker droppings to feed to its larvae. Tapeworm eggs are often present in the droppings, and can therefore find themselves being eaten by ant larvae. The parasite then has an interesting effect on the ant's behaviour when it becomes adult. It refrains from work and is fed by unparasitised workers. Parasitised ants also live longer, up to three times longer, than normal ants. This increases their chance of being eaten by a woodpecker – which benefits the tapeworm.

There are parasitic flukes who persuade their snail victim to develop a thicker shell than normal. Shells are presumably an adaptation to protect the snail and prolong its life. But a shell, like any other part of the body, is costly to make. In the personal economics of snail development, the price of thickening the shell is presumably paid out of non-shell pockets, such as those committed to reproduction. Natural selection of snails has built up a delicate balance between survival and reproduction. Too thin a shell jeopardises survival. Too thick a shell, although good for survival, takes economic resources away from reproduction. The fluke, not being a vertically transmitted parasite, 'cares nothing' for snail reproduction. It 'wants' the snail to shift its priorities towards individual survival. Hence, I suggest, the thickened shell. In extended phenotype language, natural selection favours genes in the fluke that exert a phenotypic effect on the snail, upsetting its carefully poised balance. The thickening of the shell is an extended phenotype of fluke genes, benefiting them but not the snail's own genes. This case is interesting as an example of a parasite apparently – but only apparently – doing its host a good turn. It strengthens the snail's armour and perhaps prolongs its life. But if that were really good for the snail, the snail would do it anyway, without the 'help' of a parasite. The snail balances a finely judged internal economy. Too lavish spending on survival impoverishes reproduction. The parasite unbalances the snail's economy, pushing it too far in the direction of survival at the expense of reproduction.

According to the gene's-eye view of life that I advocate, genes take whatever steps are necessary to propagate themselves into the distant future. In the case of 'own' vertically transmitted genes, the steps taken are phenotypic effects on the form, workings, and behaviour of 'own' bodies. Genes take those steps because they inherit the qualities of an unbroken, vertically travelling line of successful genes that took the same steps through the ancestral past – that is precisely why they still exist in the present. All of our 'own' genes are good companions that agree with each other about what the best steps are. Everything that helps one member of the genetic cartel into the next generation automatically helps all the others. All 'agree' about the goal of whatever it is they variously do to affect the phenotype. And why do they agree? Precisely because, in every generation, they share with each other the same exit route into the next generation. That exit route is the gametes – the sperms and eggs – of the present generation. And now we return to verticobacters and other verticoparasites. They have exactly the same exit route as the host's own genes, and therefore exactly the same interests at heart.

The genes of a verticobacter look back at the same history of ancestral bodies as its host's own genes. Verticobacter genes have the same reason to behave as good companions towards our own genes as our own genes have towards each other. If an animal benefits from fast-running legs and efficient lungs for running, then its internal verticobacters will also benefit from the same things. If a verticobacter has an extended phenotypic effect on running speed, that effect will be favoured only if it is positive from the organism's point of view too. The interests of host and bacterium coincide in every particular. A horizontobacter, on the other hand, might be more likely to 'want' its victim, when pursued, to cough with exhaustion – coughing being exactly what the horizontobacter needs in order to get itself passed on to another victim. Or another horizontobacter might want its victim to mate more promiscuously than the optimum 'desired' by the host's own genes, thereby maximising contact with another host, and hence opportunities for infection. An extreme horizontobacter might devour the host's tissues completely, reducing it to a bag of spores which eventually bursts, scattering them to the winds, where they may find fresh hosts to conquer.

A verticobacter 'wants' its victims to reproduce successfully (which means, as we saw earlier, that 'victims' is not really an appropriate word). Its 'hopes' for the future precisely coincide with those of its host. Its genes cooperate with those of the host to build a strong body surviving to reproductive age. Its genes help to endow the host with whatever it takes to survive and reproduce; with skill in building a nest, diligence in gathering food for the infants, success in fledging them at the right time to prepare to reproduce the next generation, and so on. If a verticobacter happens to have an extended phenotypic effect on a host bird's plumage, natural selection could favour verticobacter genes that brighten the feathers to make the host more attractive to the opposite sex. Verticobacter genes and host genes will 'agree' in every respect.

Exactly the same argument applies to viruses, of course. And now we approach the twist in the tail of this chapter and this book. Any virus that travels from human (for example) generation to generation via our sperms or eggs will have the same 'interests' as our 'own' genes. Whatever colour, shape, behaviour, biochemistry is best for our 'own' genes will also be best for (let's call them) verticoviruses. Verticovirus genes will become good companions of our own genes, accounting for the familiar fact that viruses can help us as well as harm us. Horizontovirus genes, by contrast, don't care if they kill their victims, so long as they get passed on to new victims by their route of choice – coughing, sneezing, handshaking, kissing, sexual intercourse, whatever it is.

A good example of a horizontovirus is the rabies virus. It is transmitted via the foaming saliva of its victims, whom it induces to bite other animals thereby infecting their blood. It also leads its victims, for example 'mad' dogs, to roam far and wide (and out in the midday sun), rather than stay, perhaps sleeping, within their normal home range. This helps the virus by spreading it over a larger geographical area.

What would be a good real example of a verticovirus? It has been estimated that about 8 per cent of the human genome actually consists of viral genes that have, over the millions of years, become incorporated. Among these 'retroviruses', some are inert but others have effects that are beneficial. For example, it has been suggested that the evolutionary origin of the mammalian placenta was the result of a beneficial cooperation with an 'endogenous' retrovirus that succeeded in writing itself into the nuclear DNA. LP Villarreal, a leading virologist, has gone so far as to suggest that 'viruses were involved in most all major transitions of host biology in evolution', and 'From the origin of life to the evolution of humans, viruses seem to have been involved … So powerful and ancient are viruses, that I would summarize their role in life as *"Ex virus omnia"* (from virus everything).'

And now, can you see where I am finally going in this chapter? In what sense are our 'own' genes different from benign, good companion viruses? Why not push to the ultimate *reductio*? Why not see the entire genome as a huge colony of symbiotic verticoviruses? This is not a factual contribution to the science of virology. Nothing so ambitious. It's more like an expansion of what we might mean by 'virus' – rather as 'extended phenotype' was an expansion of what we might mean by 'phenotype'. Our 'own' genes are verticoviruses, good companions held together and cooperating because they share the same exit route to the next generation. They cooperate in the shared enterprise of building a body whose purpose is to pass them on. Viruses as we normally understand the word, and computer viruses, are algorithms that say 'Duplicate me'. An elephant's 'own' genes are algorithms that say, in the words of an earlier book of mine, 'Duplicate me by the roundabout route of building an elephant first'. They are algorithms that work only in the presence of the other genes in the gene pool. They are equivalent to an immense society of cooperating viruses.

I'm not just saying that our genome consists of 'endogenous retroviruses' (ERVs) that were once free, infected us, and then became incorporated into the chromosomes. That is true in some cases and it is important, but it's not what this final chapter is suggesting. Lewis Thomas also didn't mean what I now mean, although I would love to borrow his poetic vision in pushing the climax of my book.

> We live in a dancing matrix of viruses; they dart, rather like bees, from organism to organism, from plant to insect to mammal to me and back again, and into the sea, tugging along pieces of this genome, strings of genes from that, transplanting grafts of DNA, passing around heredity as though at a great party.

The phenomenon of 'jumping genes', too, is congenial to my vision of a genome as a cooperative of verticoviruses. Barbara McClintock won a Nobel Prize for her discovery of these 'mobile genetic elements'. Genes don't always hold their place on a particular chromosome. They can detach themselves, then splice themselves in at a distant place in the genome. Some 44 per cent of the human genome consists of such jumping genes or 'transposons'. McClintock's discovery of jumping genes conjures a vision of the genome as a society, like an ants' nest: a society of viruses held together only by their shared exit route, and hence shared future and shared actions calculated to secure it.

My suggestion is that the important distinction we need to make is not 'own' versus 'alien' but *vertico* versus *horizonto*. What we normally call viruses – HIV, coronaviruses, influenza, measles, smallpox, chickenpox, Rubella, rabies – are all *horizonto* viruses. That, precisely, is why many of them have evolved in a direction that damages us. They pass from body to body, via routes that are all their own, by touch, in the breath, by genital contact, in saliva, or whatever it is, and not via the gametic routes with which our own genes traverse the generations. Viruses that share the same genetic destiny as our own genes have no reason to dissent from good companionship. On the contrary. They stand to gain from the survival and successful reproduction of every shared body they inhabit, in exactly the same way as our own genes do. They deserve to be considered 'our own' in an even more intimate sense than mitochondria, for mitochondria pass down the female line only. And, from this point of view, our 'own' genes are no more 'own' than a retrovirus that has become incorporated into one of our chromosomes and stands to be passed on to the next generation by exactly the same sperm or egg route as any other genes in the chromosome.

I cannot emphasise strongly enough that I am *not* suggesting that all our genes were once independent viruses that later 'came in from the cold' and, as retroviruses, 'joined the club' of our own nuclear genome. That is known of some 8 percent of our genes, it may be true of many more, it is interesting and important, but it is not what I am talking about here. My point is rather to downplay the distinction between 'own' and 'other', and to emphasise instead the distinction between vertico and horizonto.

Our entire genome – more, the entire gene pool of any species of animal – is a swarming colony of symbiotic verticoviruses. Once again, I'm not talking only about the 8 percent of our genome that consists of actual retroviruses, but the other 92 percent as well. They are good companions precisely because they are vertically transmitted, and have been for countless generations. This is the radical conclusion towards which this chapter has been directed. The gene pool of a species, including our own, is a gigantic colony of viruses, each hell-bent on travelling to the future. They cooperate with one another in the enterprise of building bodies because successive, temporary, reproduce-and-then-die bodies have proved to be the best vehicles in which to undertake their vertical Great Trek through time. You are the incarnation of a great, seething, scrambling, time-travelling cooperative of viruses.

Endnotes

Some readers enjoy reflective 'by the way' digressions. Others decry them as intrusive distractions from the point at hand. In this book I have bowed to the decriers, banishing my digressions to endnotes, and even going so far as to remove all endnote reference numbers from the text. Instead, the endnotes are anchored by page numbers and identifying sentences. And turning to those who enjoy 'by the way' reflections, perhaps they might read some endnotes for amusement in their own right, regardless of the point from which they digress.

Chapter 1. Reading the Animal

P 2. *I arbitrarily assume SOF to be female.* Gendered pronouns are almost guaranteed to offend somebody. I'm not fond of laboured constructions like, 'He or she must ask himself or herself whether it is fair to the reader to subject him or her (or them) to such tortured language'. In many languages, even the word 'reader' might have to be doubled up: '*Leser oder Leserin?*' I favour an alternative convention, of a piece with hand-kissing courtesy, or old-world curtsying, whereby authors adopt the gender pronouns opposite to their own sex. Since I happen to be male, I refer to my hypothetical future scientist as 'she'. If I were a female author, I'd do the reverse.

P 3. *Professor Hamilton is widely regarded as the most distinguished Darwinian of his generation.* See his experiment in autobiography, *Narrow Roads of Gene Land* (3 volumes), in which his technical scientific papers are reprinted between soul-baring essays in personal memoir. 'Therefore one last confession. I, too, am probably coward enough to give funds for "elixir" gerontology if anyone could persuade me that there is hope. At the same time I want there to be none so that I will not be tempted. Elixirs seem to me an anti-eugenical aspiration of the worst kind and to be no way to create a world our descendants can enjoy. Thus thinking, I grimace, rub two unrequestedly bushy eyebrows with the ball of a happily still-opposable thumb, snort through nostrils that each day more resemble the horse-hair bursts of an old Edwardian sofa, and, with my knuckles not yet touching the ground, though nearly, galumph onwards to my next paper.'

P 3. *Robert Trivers, mourning his death.* Trivers (2000).

P 3. *'Through a glass darkly'.* 1 Corinthians 13, 12. This endnote should have been unnecessary. The phrase is one of the 129 biblical phrases I listed in *The God Delusion* as being, like many of Shakespeare's, an essential part of our culture, necessary equipment for a full, literate life. I was arguing in favour of religious education, meaning education about religion rather than indoctrination in a particular religion.

P 5. *Variant individuals whose predictions are slightly wrong, or slightly more wrong than those of competitors within their own species. That is natural selection.* Darwinian natural selection chooses within populations, not between them. Darwin was very clear about this (with one exception when talking about humans in *The Descent of Man*, 1871). The dinosaurs were replaced by mammals, but that replacement was not a Darwinian selective event. What was Darwinian was the differential survival of individual mammals within each mammal species, reflecting individual success in the enterprise of filling the shoes of some particular extinct dinosaur species.

P 6. *Sense organs do not faithfully project a movie of the outer world into a little cinema in the brain.* There's no little homunculus sitting in a 'Cartesian Theatre' (Dennett, 1991).

P 7. *The frog's eye largely confines itself to reporting small moving objects to the brain.* Lettvin et al. (1959) went on to measure nerve impules in single neurons in the brain of the frog when they presented things to the eyes. The 'sameness' neuron, for example, is initially silent when a small moving object is presented. Then, suddenly, it 'notices' the object and starts firing. The firing rate inccreases every time the object changes its pattern of movement, for example turns a corner.

P 7. *The eye's wiring-up embodies a prediction … likely to hit food.* The physiologist Horace Barlow (1961, 1963), who did pioneering work on the frog visual system, had an inspiring vision of sensory systems in general, which is very congenial to this book. He summed up neural representations as 'approximate estimates of the probable truths about the current environment'. He didn't put it in these words, but his idea was equivalent to this: the way an animal's sensory system is tuned is a kind of negative rendering of the statistical properties of its world.

P 7. *Can be broken down into component sine waves – 'Fourier analysis'.* Once, in the Kruger National Park, I came upon the urine trail that a male elephant in musth had made in the dust. It looked approximately sinusoidal, and had evidently been made by his dribbling penis swinging as a pendulum. I photographed it with a vague idea of getting a mathematician to Fourier analyse it and compute the length of his penis. Like so many of my schemes, it never came to anything.

P 8. *Today's equivalent of my grandfather would use a computer screen instead of a clothesline.* Computer simulations provide an especially useful tool for modelling. Nothing really goes on inside a computer except extremely rapid shuttling of billions of 0s and 1s. But the readout from a computer can represent a chess game, the world's weather, traffic patterns in Birmingham's Spaghetti Junction, a pendulum, the Eroica symphony, population cycles of lemmings and arctic foxes, the city of Vancouver, or waves of muscle fibre contraction in a heart.

P 10. *Svante Pääbo won a Nobel prize for jigsawing the genome of Neanderthal and Denisovan humans.* In my opinion, Pääbo and his colleagues should be honoured also for their disciplined and scrupulous working-out of the methods by which research in ancient DNA should be carried out. The pitfalls are many and deep, above all contamination by modern DNA. Failure to appreciate this has led to premature, headline-grabbing reports from around the world, of absurdly ancient DNA.

P 10. *Other possible 'resurrections' might include the dodo, passenger pigeon, great auk, and thylacine (Tasmanian wolf).* Reviving extinct mammals raises the problem of finding a substitute womb among living animals. The DNA of Steller's sea cows has been recovered and sequenced. It would be wonderful to think that they might be brought back to life. A probably insurmountable problem would lie in finding a surrogate mother in which to implant an embryo. Their surviving cousins, the dugongs and manatees, are too small to give birth to them. In this respect, mammoths are a better bet, for surviving elephants are big enough to gestate them. Interestingly, the problem doesn't arise with marsupials because they are born extremely small, then crawl into the pouch. A substitute pouch is much easier to improvise than a substitute womb. A Labrador-sized Tasmanian wolf, *Thylacinus*, would have been the size of a rice grain at birth, not very different from a newborn dunnart, its mouse-sized surviving relative. Embryologists of the far future will probably be able to grow embryos outside the womb. It is the beauty of DNA's digital nature that there is no need to preserve any actual biological material. Those future embryologists need only go to the library and download the genome.

P 11. *The distribution of both genes and words across the Micronesian islands.* Cavalli-Sforza & Feldman, 1981.

P 11. *Sir D'Arcy Thompson (1860–1948), that immensely learned zoologist, classicist, and mathematician. On Growth and Form* (1942), his beautifully written *magnum opus*, is laced with quotations in many languages including Latin, Greek, French, German, Italian and Provençal, usually without translation except in the case of Provençal, which he did deign to translate for us (into French). I thank Denis Noble for confirming the dialect of the Provençal quotation, which is from a poem by the great naturalist Jean-Henri Fabre. It prompts from Thompson a poignant aside on an old man's struggle against gravity. 'But in part, the slow decline in stature is an expression of an unequal contest between our bodily powers and the unchanging force of gravity, which draws us down when we would fain rise up. For against gravity we fight all our days, in every movement of our limbs, in every beat of our hearts; it is the indomitable force that defeats us in the end, that lays us on our deathbed, that lowers us to the grave.' I said 'poignant' because, in his prime, Sir D'Arcy had been, to quote Peter Medawar, 'over six feet tall, with the build and carriage of a Viking, and with the pride of bearing that comes from good looks known to be possessed'.

P 12. *Once fertilisation has taken place, the genome remains fixed.* Well, almost entirely fixed. In some respects we need that 'almost', but it isn't important for the point being made in this chapter. 'Somatic' mutations, mutations of cells within the body, do occur and they may spawn lineages of cells evolving within the body. These, with subsequent somatic mutations, can become tumours. Natural selection of rebel cells within the body then can turn tumours malignant. Later, it can make them better at being malignant – but the opposite of better for the body as a whole. We shall return to this in Chapter 12. For purposes of the chapter, we are concerned only with mutations in the so-called *germline*, that is mutations that may be passed on to future generations.

P 12. *It is the gene pool of the species, not the genome of the individual, that changes under the Darwinian chisel.* It is all too easy to confuse evolution with development. In development, a single entity changes. In evolution, each of a series of entities is slightly different from its predecessor, like successive frames of a movie film. Astronomers are wrong to talk of stars 'evolving' through a 'main sequence', as in 'The Sun's evolution will culminate in a red dwarf.' There are not successive generations of suns. The one sun changes. No, the Sun's *development* will culminate in a red dwarf. We see the shape of a living body *develop* via the process of embryology. We see the shape of a body *evolve* when we arrange successive generations in series and observe how the typical shape changes from generation to generation. The shape of a typical member of a species *evolves* while gene pool *develops*.

P 13. *Progressive changes in average phenotype.* Biologists in progressive circles are supposed to look askance at the idea of evolution being progressive. Rightly so, in the case of the popular caricature of evolution striving towards the alleged climax of *Homo sapiens*. But when we are talking about the evolution of complex organs like eyes, you can't get away from the idea of progressive improvement. A fully functioning vertebrate eye must have passed through less efficient intermediate stages in an escalating series, as a matter of logical necessity. A good account of the history of ideas of progress in evolution is given by Michael Ruse (2010).

Chapter 2. 'Paintings' and 'Statues'

P 15. *Here's another highly camouflaged lizard.* Photo by kind permission of Yathin Krishnappa. This particular species of lizard has a second trick up its sleeve. It's a *Draco* 'flying lizard', better called gliding lizard. If a predator were to break the camouflage, the *Draco* would launch itself into the air, and glide gracefully to another tree a safe distance away. Its 'wings' are not modified arms like those of a bird or a bat. Instead, it spreads its ribs out sideways, and a membrane of skin between the ribs makes an excellent surface to catch the air.

P 19. *Dark, melanic mutant of the same species of moth that you may have noticed less immediately by its side.* In a rural area, it is the dark form that is conspicuous, the light one camouflaged. The maintenance of the two morphs in the same species ('polymorpism') is maintained by opposite selection pressures in the two areas since the industrial revolution, as convincingly shown by Bernard Kettlewell (1973). Thereby hangs a fascinating tale, but it would take us too far afield.

P 21. *It does it by means of a 'statue', which fools the fish.* Barnhart et al. (2008).

P 21. *This highly camouflaged snake from Iran has a dummy spider at the tip of its tail.* https.//www.mirror.co.uk/news/world-news/incredible-snake-us-es-tail-looks-5971693. https.//www.youtube.com/watch?v=XFjoqyVRmOU

P 22. *Where are the snows of yesteryear? Mais où sont les neiges d'antan?* Rosetti's translation of François Villon's famous line.

P 23. *Vietnamese mossy frog.* Photograph kindly supplied by Michael Sweet, as is the yellow striped beetle on the next page.

P 27. *Black-headed gulls respond to a model gull head on the end of a stick, as though it were a whole real gull.* RF Mash, quoted in Tinbergen (1964).

P 27. *Eyes painted on the rumps of cattle are effective in deterring predation by lions.* Radford et al. (2020). Photo by Cameron Radford, kindly supplied by Neil Jordan.

P 27. *We could call it the Babar effect.* De Brunhoff (1935).

P 28. *All I can say is that if I were a would-be predator, I wouldn't hang about* (Crew, 2014).

P 28. *The octopus.* Still from a movie by Michael Sweet.

P 28. *The vulture.* Photo by Hussein Latif.

P 29. *Some butterflies have a false head at the back of the wings.* Hendrik et al. (2022).

Chapter 3. In the Depths of the Palimpsest

P 32. *Every twinge of lower-back pain reminds us that our ancestors less than 6 million years ago walked on all fours.* Lents (2019).

P 32–33. *That salty, larger-than-life intellectual warrior JBS Haldane.* Haldane (1940). Stories about him are legion. As a front-line officer in the First World War he once rode a bicycle across a gap, in full view of the Germans, to prove his confidence – in the event justified – that they'd be too astonished to shoot. Like his slightly less famous father, he did dangerous experiments on himself. One minor consequence was a punctured ear drum, and he then developed a parlour trick, when smoking his pipe, of puffing smoke out of his ear.

P 33. *The lobefins … crawled out of the sea.* Lobefins that stayed behind include the coelacanths. They actually went the other way, into the deep sea, which is perhaps how they managed to escape extinction. They were thought to have gone extinct along with the dinosaurs until Marjorie Courtenay-Latimer, museum curator, spotted one in 1938, in the catch of a South African trawler. Incredulous as to its true identity, she summoned a leading fish expert, Professor JLB Smith (1956), who gave a spine-tingling account of his first sight of it. 'I would not have been more surprised if I had seen a dinosaur walking down the street … that first sight hit me like a white-hot blast and made me feel shaky and queer, my body tingled. I stood as if stricken to stone. Yes, there was not a shadow of doubt, scale by scale, bone by bone, fin by fin, it was a true Coelacanth. It could have been one of those creatures of 200 million years ago come alive again. I forgot everything else and just looked and looked, and then almost fearfully went close up and touched and stroked while my wife watched in silence.' He formally named the fish *Latimeria* after its discoverer.

P 34. *The general name for learned behaviour becoming genetically incorporated is the Baldwin Effect.* An unusually intelligent individual discovers a clever trick and learns to perfect it. Perhaps others imitate it, as when blue tits learned to open milk bottles on British doorsteps and the habit radiated out across the country as a memetic epidemic. Milk is no longer delivered on doorsteps, but if the practice had gone on for a long time, the Baldwin Effect might have kicked in. Individual tits genetically equipped to learn the habit fastest would have passed their genes on disproportionately. Eventually, birds would evolve that learn so fast they didn't need to learn at all. The habit, originally learned, would have been genetically assimilated.

P 35. *When this evidence was called into question, Romer's whole theory suffered in appreciation.* I never stopped thinking it was a good idea, and was pleased when the astronomer Steven Balbus gave Romer's theory (and his fish) new legs. Earth's unusually large moon generates, with the sun, high-amplitude tidal changes in sea level. Tide pools are left high and drying. In Devonian times the moon was twice as close as it is today, and tidal amplitudes even greater. Fish would frequently have found themselves stranded in drying pools by the receding tide. Any ability to cross over to neighbouring deeper pools would have been at a premium. It is easy to see how these conditions could have provided the initial impetus to the evolution of adaptations to land life. Balbus's refinement of the Romer theory (there's *much* more to it than my extremely brief, non-mathematical summary) doesn't need to postulate drought. It's so pleasing when an expert in a completely different field of science uses it to contribute to biology. Balbus's (2014) theory has even inspired a poem, 'Tidal Evolution', by Sam Illingworth (2020).

P 35. *Otherwise, they totally recommitted to the sea.* One genus of fish, the grunion *Leuresthes*, whose ancestors, unlike those of turtles, never lived outside the sea, moves up the beach to bury its eggs above the tideline, out of reach of marine predators. They do so *en masse*, in an astonishing spectacle known as the California Grunion Run. The young hatch two weeks later, taking the high tide at the flood and restoring themselves to the sea. Rowland (2010).

P 40. *Two nations 'separated by a common language'.* Attributed to George Bernard Shaw.

P 42. *Anyone who wants to quantitatively decipher.* Yes, it's a split infinitive. I like split infinitives, and would like them even if they lacked the imprimatur of Fowler's *Modern English Usage* (1968). They accurately convey meaning.

P 47. *If this giant Galapagos tortoise could sing a Homeric epic of its ancestors.* 'The time of the singing of birds is come, and the voice of the turtle is heard in our land' (*Song of Songs*). Please don't tell me it's a mistranslation of 'turtle dove'.

P 47. *In the genetic book of the dead, older scripts of the palimpsest can amount to 'constraints on perfection'.* This is the title of a chapter in *The Extended Phenotype*, briefly summarised in Chapter 4 of this book.

P 49. *If you think of it as design, this is obviously rotten design.* Rotten design in the human body is ubiquitous enough to fill a book, and Nathan Lents (2019) has written it.

P 49. *The length of the detour in the giant dinosaur Brachiosaurus would have been about 20 metres.* Wedel (2012).

P 50. *The marginal cost of each small lengthening of the detour.* 'Marginal cost' is a favourite phrase of economists, but what does it mean in evolution? Here it means that in any particular generation, individuals who paid the slight cost of lengthening the detour would have survived better than rival individuals who paid the major cost of a radical change to their embryology.

P 50. *There's a very similar example in the routing of the tube connecting testis to penis.* Instead of going directly, it diverts to loop around the tube connecting kidney to bladder (Williams, 1996b).

P 51. *A group of extinct South American herbivores, the Litopterns.* Simpson (1980).

P 51. *Arthropods have a different Bauplan (building plan or body plan).* The German word is customarily used by biologists. It has become sufficiently embedded in English as a loan word that I prefer to use the English plural 'bauplans' rather than the strictly correct German *Baupläne*.

P 52. *Whales sprang from within the artiodactyls, the even-toed, cloven-hoofed ungulates.* Nikaido et al. (1999).

P 52. *A staggering revelation that nobody saw coming.* Actually, and amazingly, that's not quite true. In 1866 the zoologist Ernst Haeckel, Darwin's leading German champion, published a family tree of all mammals in which he placed 'Obesa' (hippos) as the sister group of all whales. He later (1895) changed his mind. He was right the first time – like, on several occasions, his hero Charles Darwin, such that the first edition of *Origin* is scientifically more accurate than the sixth.

P 53. *They are also sometimes called 'junk' genes, but they aren't 'junk' in the sense of being meaningless.* There are also repetitive sequences of DNA, which have a better claim to be 'junk' since they really are meaningless in the sense of not coding for protein chains. In any case, so-called 'junk' genes might be regarded as selfish genes in the sense of my book of that name. On page 47 of *The Selfish Gene* (1976), I wrote, 'a large fraction of the DNA is never translated into protein. From the point of view of the individual organism this seems paradoxical. If the "purpose" of DNA is to supervise the building of bodies, it is surprising … But from the point of view of the selfish genes themselves there is no paradox. The true "purpose" of DNA is to survive, no more and no less. The simplest way to explain the surplus DNA is to suppose that it is a parasite, or at best a harmless but useless passenger, hitching a ride in the survival machines created by the other DNA' (Dawkins, 1976). This suggestion was developed further by Doolittle & Sapienza (1980) and by Orgel & Crick (1980).

P 53. *They've become side-lined as pseudogenes.* I learned of this wonderful example from Jerry Coyne's (2009) book *Why Evolution Is True*, which goes into the matter in greater detail.

P 53. *They can be read in total clarity.* Even though the reading is distorted by a bombardment of uncorrected mutations, accumulated over the time span since the pseudogene ceased to be read by the organism itself.

Chapter 4. Reverse Engineering

P 56. *Lewontin defined adaptationism as 'That approach ... solutions to problems'.* Lewontin (1979).

P 57. *They are literally neutral, in the sense of the Japanese geneticist Motoo Kimura.* John Maynard Smith told a funny sorry about Kimura's (1983) book. Kimura reluctantly recognised that some characteristics really were naturally selected, Darwinian adaptations. But so reluctant was he, that he couldn't bear to write the sentence himself. So he asked his colleague, the American geneticist James Crow, to write it for him. It's a good story, typical of dear John Maynard Smith, but hard to believe. How could any intelligent person fail to be impressed by the obvious design features of living organisms?

P 58. *Almost literally the pivot of the animal's life.* The quotation from Cain (1966, reprinted 1989) continues. '*Polyxenus* is a very remarkable minute millipede which can actually walk upside-down on the ceiling of small crevices and even moult there. Manton shows that a curious Y-shaped bar of chitin on the legs enables the animals to use a very wide leg-swing in walking and develop considerable fleetness without using long legs. Speed is necessary as it has to make long journeys for its food, and short legs are an advantage in the crevices where it hides. Its gait is basically of a slow pattern, thus enabling it to have many leg-tips touching the ceiling of a crevice at once; also more secure adherence is obtained by means of special lappets at the tips. She further points out that the Y-shaped bar is also produced completely independently in some very fast-running centipedes for the same reason, namely, to strengthen the joints of a very widely-swinging leg.'

P 58. *JBS Haldane made a relevant hypothetical calculation.* Haldane (1932).

P 59. *'That is the one point, which I think all evolutionists are agreed upon, that it is virtually impossible to do a better job than an organism is doing in its own environment'.* Lewontin (1967).

P 60. *Reverse engineering is the method by which we can reconstruct the purpose of the antikythera.* Gregory (1981).

P 61. *John Krebs and I revived the idea of the evolutionary arms race.* Dawkins & Krebs (1979).

P 63. *'From such warped beginnings nothing debonair can come'.* Cecil Day-Lewis, *The Unwanted.* 'Willy nilly born it was, divinely formed and fair'.

P 66. *Brontosaurus.* The rules of biological nomenclature dictate that earlier namings take precedence. Readers may be aware that the name *Brontosaurus* was replaced by *Apatosaurus* for this reason. The name *Brontosaurus* was revived when it was decided that there were actually two genera. It is a pleasure to return to a childhood favourite. Gould (1991), Callaway (2015).

P 68. *I have plotted mean arterial blood pressure against the logarithm of body mass for a range of mammals from mouse (Mus musculus) to elephant (Loxodonta africana).* The data from which I plotted the graph are from

Poulsen et al. (2018). The straight line is a simple linear regression. To be fair, I should add that the purpose of Poulsen et al.'s paper is to cast doubt on the validity of the previously published data, mainly on the grounds that the conditions for measuring blood pressure were too variable. But that should cast doubt on the upward trend with body size only if measuring methods varied *systematically* with size, and there seems no reason to suppose this. The high figure for the giraffe (*Giraffa camelopardus*) is pretty convincing, regardless of their misgivings.

P 68. *It's best to use logarithms for the weights – otherwise it would be hard to fit mouse and elephant on the same page.* There are other reasons for using logarithms. See endnote on page 310.

P 68. *Surprisingly, other evidence shows that the giraffe heart is not especially large.* Østergaard et al. (2013).

P 69. *They seem designed to slice rather than to mill.* 'Molar' is a poor name in the case of a carnivore, for it comes from the Latin *mola*, millstone, and the molars of carnivores are the very opposite of millstones. The name comes from human anatomy where the molars really look like, and function as, grinding millstones. An extreme example of the same kind of human-centred misnaming concerns a bone in the jaw of fish with an absurdly long name. We humans have separate bones called the palatine, the pterygoid, and the quadrate. To back-combine these in a single bone in the fish jaw called the palatopterygoquadrate seems to give vivid new meaning to 'putting the cart before the horse'.

P 70. *They have thick walls toughened by cellulose and silica.* The great American palaeontologist GG Simpson (1953) argued that early horses were divided between browsers (feeding on tree leaves; these horses are now extinct) and grazers (on grass). Grazers typically have higher-crowned, more complicated teeth, probably to cope with the silica-rich cells of grasses. It's been suggested that the high-crowned grinding teeth of horses evolved in lockstep with the evolution of grasses. However, there is evidence (Strömberg, 2006) that grasses arose well before high-crowned horse teeth, so 'lock-step' may be the wrong way to put it.

P 70. *A plausible reason [for cat family running to long canines more than dog family] is as follows.* It was suggested to me by Gary Meaney (2022), a young zoologist and author from Dublin.

P 71. *'Heated in the chase'.* 'As pants the hart for cooling streams / When heated in the chase'. Hymn (1696), based on Psalm 42, 'As the hart panteth after the water brooks'.

P 74. *Their prey is small enough to swallow whole.* Dolphins normally swallow their fish whole, but there are fascinating stories of some dolphins biting, but not swallowing, a puffer fish as a recreational drug. Puffer fish release a neurotoxin as a defence mechanism when attacked. In high doses it is lethal but in low doses it has a mild narcotic effect, which incidentally appeals to Japanese sashimi gourmets. Chefs undergo rigorous training in the art of removing the lethally poisonous parts. A BBC documentary shows dolphins chewing on a puffer fish and passing it around between them, like people sharing a spliff. The dolphins appear to enter a trance-like state as a result. The puffer fish swims away at the end, presumably somewhat damaged but not swallowed. BBC1, *Dolphins – Spy in the Pod*, Episode 2. Clip 'Pass the Puffer', at https.//www.youtube.com/watch?v=msx3BAhIeQg.

P 77. *'Anteater' is not a great name for a taxonomic group.* Equally unhelpful are 'Carnivora' (other mammals eat meat) and 'Insectivora' (other mammals eat insects. I'm happy to say that this last name has recently been officially discontinued).

P 77. *The ancient supercontinent of Gondwana.* From before the Cambrian until the Jurassic, Africa, Madagascar, South America, Antarctica, Australia, and New Zealand were united in a giant continent called Gondwana, or Gondwanaland.

P 78. *Huge numbers of the insects stick to the [Giant Anteater's] tongue.* Van der Linden (2016).

P 81. *The intriguing hypothesis that the invention of cooking was the key to human uniqueness and human success.* Wrangham (2009).

P 84. *A beautiful and still proceeding long-term study of 'Darwin's Finches' on one of the smaller Galapagos Islands by Peter and Rosemary Grant.* Grant & Grant (2014), Weiner (1994).

P 84. *The evolutionary divergence of Hawaiian honeycreepers shows a variety of beaks that outdoes even that of the Galapagos finches.* Pratt (2005).

P 85. *The range of [Hawaiian honeycreeper] bill types that has evolved in such a short time is astonishing.* To be fair, the time available is actually rather longer than the 5 million years that is the age of the oldest island, Kauai. This is because the archipelago itself is older, including islands that would once have been populated by birds but have now sunk beneath the waves. Similar considerations apply in Galapagos.

Chapter 5. Common Problem, Common Solution

P 88. *Matt Ridley's How Innovation Works documents …* Ridley (2020).

P 89. *A captive specimen called Benjamin survived in Hobart Zoo until 1936.* There's a sad movie (coloured from an original black and white) of him in his cage. Search for YouTube Thylacine.

P 89. *Stag beetles, but not stags, can lift their rivals high in the air on the prongs of their 'antlers'.* The weight of an animal increases as the cube of its size. The strength of a muscle, on the other hand, will tend to increase only as the square of its size, because the number of muscle fibres acting in parallel is proportional to the muscle's cross-sectional area. Stags being much larger than stag beetles, it takes disproportionately more muscular strength for a stag than for a beetle to lift a rival off the ground. The whole topic of how things scale with size is comprehensively treated by Geoffrey West (2017), a splendid example of the contribution a clever physicist can make to biology. The span of West's insights is breathtaking, encompassing everything from bacteria to cities within the same mathematics.

P 93. *A pangolin … formed itself into a ball and rolled down the slope.* Tenaza (1975).

P 94. *Eyes have been independently evolved many dozens of times, to nine basic designs.* Land (1980).

P 95. *New World quills prolong the agony by means of backward-facing barbs.* Perkins (2012).

P 95. *The barbs of bee stings (American stingers).* In the case of the worker bee sting, the barbs prove fatal to the bee itself. The very property that makes it such an effective weapon makes the sting almost impossible for the bee to withdraw. The bee is stuck to its victim, and when the victim brushes it off, the sting remains embedded, together with some vital internal organs which are torn out of the doomed bee's body. The disembodied venom-pump continues to inject its poison into the victim even after being torn out of the bee itself. It is kamikaze, a suicide attack, beautifully compatible with the gene's-eye view of natural selection. Worker bees are sterile. Selection favours genes that programme workers to commit suicide for the benefit of copies of the same genes passed on by reproductive members of the hive, queens, and males. If given enough time, a bee can theoretically save her life by 'unscrewing' herself, walking round in circles. As a boy, I once witnessed a bee unscrewing itself after stinging me on the hand. My proud report (Dawkins, 2013) of that youthfully altruistic forbearance was disputed by a beekeeper (Garvey, 2014) who denied that bees can unscrew themselves. I was vindicated in a film, which also contains a good account, with animation, of the alternate sawing action of the two serrated blades (https.//www.youtube.com/watch?v=nTVsqc2CCGo).

P 96. *The stinging cells, 'cnidoblasts', of jellyfish are miniature harpoons that shoot out on the ends of threads, and inject venom.* They give their name to the phylum 'Cnidaria' to which jellyfish, hydras, corals, and sea anemones belong. 'Cnide' means 'nettle' in Greek. Sea anemones also have cnidoblasts, which they use to harpoon small prey. If you touch a sea anemone's tentacle, it seems to cling to you. Your finger is being speared by hundreds of tiny harpoons, each still attached to its cnidoblast. Unlike jellyfish, most sea anemone cnidoblasts don't have enough venom to hurt you, but they can cause a rash.

P 98. *Also very mole-like, among the rodents there are the zokors of Asia.* Simon Conway Morris (2003) lists thirteen convergent 'moles' from all around the world.

P 103. *Using echoes of their own sounds to detect targets.* Donald Griffin (1959) was the main discoverer of bat echolocation, a term that he coined. He also had a strong interest in the puzzle of how migrating birds navigate. There's an apocryphal story that he had a van, fully equipped with instruments, in which he would make unannounced spot-checks on research labs working on magnetic orientation in birds.

P 104. *Some megachiropterans use a less precise form of echolocation, but they generate their clicks with their wings rather than with the voice.* Boonman et al. (2014).

P 104. *Interestingly, molecular evidence indicates that one group of Microchiroptera, the Rhinolophids, are more closely related to Megachiroptera than they are to other Microchiroptera.* Teeling et al. (2000).

P 105. *Thomas Nagel didactically asked, 'What is it like to be a bat?'* Nagel (1974).

P 106. *Dolphins pay special attention to pregnant women swimming with them.* Gallagher (2020).

P 107. *If you construct a tree of resemblance based on prestin sequences alone – something remarkable emerges. Dolphins and small bats cluster together with each other.* Li et al. (2010).

P 107. *Another example concerns flight surfaces in mammals.* Feigin et al. (2023).

P 108. *There are probably lots of shared features buried deep in an aquatic animal's physiology and genome. We have just to find them.* Huelsmann et al. (2019).

P 108. *What you do for a genome-wide association study.* Francis Collins, personal communication.

P 111. *What they studied was not aquaticness but hairlessness in mammals.* Kowalczyk et al. (2022).

P 113. *There is a long-extinct trilobite that also had a huge paddle-like appendage like that of the paddlefish … yet another convergence.* S Turvey, quoted in Dawkins & Wong (2016), a book which also discusses the platypus and paddlefish electric senses.

P 116. *The cleaner trade is practised by sixteen different species, having evolved within the Palaemonidae five times independently.* Horka et al. (2018).

P 116. *Simon Conway Morris has treated convergent evolution more vividly and thoroughly than anyone else.* Conway Morris (2003, 2015).

Chapter 6. Variations on a Theme

P 121. *The fish is able to adjust its specific gravity and achieve perfect equilibrium at whatever happens to be its preferred depth at any time.* A fish with a swim bladder is a sophisticated form of Cartesian Diver. Cartesian Diver is the name given to a simple toy containing an air bubble. You place the Diver in a bottle of water. The air bubble shrinks when the water pressure in the bottle increases, and this changes its buoyancy equilibrium position. If properly adjusted (it isn't difficult) so that it is in equilibrium in the middle of the bottle, you can sensitively raise or lower that preferred point by gently pressing or releasing the cork. A stopper that screws into the top of the bottle (in Britain I think of large cider bottles) allows even more precision control. It causes the bubble in the Diver to shrink or expand, like a fish with its swim bladder. The fish is in control of its swim bladder, not some outsider manipulating a cork, but otherwise it is a Cartesian Diver. When I saw manatees in Florida I was impressed by their dreamy fish-like drifting, but they don't have a swim bladder. It has been shown that their centre of gravity is shifted forward to compensate for a natural tendency to tilt head-up – a disadvantage in an animal that grazes on the bottom. And they have heavy bones to compensate for what would normally be a tendency to float to the surface. But when they do need to surface, how do they do it? It turns out that they adjust their point of hydrostatic equilibrium by controlling the volume of the lungs. A pelagic (swimming near the sea surface) octopus called *Oxythoe* has convergently evolved a swim bladder. Packard & Wurtz (1994).

P 124. *There are about 400 species of Cichlid (pronounced 'sicklid') fish in Lake Victoria.* Estimates of species numbers vary quite widely. My authority is George Barlow's book (2000).

P 125. *This hunting technique [predatory fish pretending to be dead to lure prey] was thought to be unique in the animal kingdom.* Barlow (2000). 'Playing possum', pretending to be dead to avoid being attacked, is more common.

P 125. *My old friend the late George Barlow vividly described the three great lakes of Africa as Cichlid factories. His book makes fascinating reading.* Barlow (2000).

P 128. *Selection pressures have actually been measured in the wild, for example on butterflies.* Ford (1975).

P 130. *Julian Huxley applied D'Arcy Thompson's method to the relative growth of different body parts in the developing embryo.* Huxley's book (1932) initiated a major field of research on relative growth rates of different body parts. As an animal grows, its various parts normally don't grow at the same rate, which would be called isometry. Allometric (disproportionate) growth usually follows a mathematically lawful relation in which the growth rate of one bit of the animal is proportional to the growth rate of another bit raised to a power. Evolutionary change may then show itself as a genetically caused change in the power, or the constant of proportionality, or both. The mathematically lawful differences between adult body forms, as diagrammed by D'Arcy Thompson, are then explained evolutionarily as mathematically lawful changes in growth rates in different parts of the body. Such changes are under genetic control, as different genes are switched on in the cells of different parts of the embryo. If the power term in the equation remains fixed in a group of animals, say mammals, a mathematical consequence is that you can plot a scatter diagram of the logarithm of size of one body part against the logarithm of size of the other body part, and the points will fall along a straight line. You can then ask why some points lie above the line, others lie below it. For example, monkeys have bigger brains than is normal for mammals of their body size.

P 132. *The energy released [by the blow of a mantis shrimp's clubs] is so great that the water boils locally, and there is a flash of light.* This is caused by cavitation, a short time after the hammer blow itself. Patek (2015).

P 132. *You don't want to mess with a mantis shrimp* Patek (2015).

P 132. *(Literally) stunning 'pistol shrimps' or 'snapping shrimps'.* Kaji et al. (2018).

P 132. *These have one enlarged claw, somewhat bigger than the other.* In this respect they resemble the much more asymmetrical fiddler crabs, who use their one hugely enlarged claw – left- and right-handers are equally common – to semaphore to one another, each species waving the claw with its own characteristic rhythm.

P 134. *The genes controlling the segments are arrayed in series along the length of a chromosome.* Deutsch & Mouchel-Vielh (2003).

P 134. *The cold stream on our farm, in the course of which my parents dug out a pool for us to swim.* Microsoft Word's language checker tried to change 'in the course of' to 'during', and you have to sympathise. But this was a rare opportunity to use 'in the course of' in its literal rather than metaphorical sense. An opportunity not to be missed, although I admit that the metaphorical usage has become so common as to almost obliterate the original. Scholars such as CS Lewis (1939) have noted that many, even most, of our words are metaphorical transformations from earlier meanings. 'Inspiration' used to mean 'breathing in', for instance. 'Course' comes from the Latin for run, hence the course in which a river runs, or, the course along which an argument runs.

P 134. *Patel and his colleagues.* The work of the Patel group is published in a series of papers, usefully summarised in Shubin (2020).

P 140. *Presumably parasites, although nobody knows who their victims are.* Glenner et al. (2008).

P 141. *See Darwin's own drawing.* Deutsch (2010). For some eight years, during which he might have been publishing his great theory of evolution, Darwin devoted himself to writing several detailed monographs on barnacles. It is said that when one of Darwin's children was being shown around the house of a friend, he asked, 'But where does your papa do his barnacles?', it being assumed that 'doing barnacles' was how any father must spend his time.

P 142. *Ruder heads stand amazed ...* Sir Thomas Browne (*Religio Medici*, 1643).

Chapter 7. In Living Memory

P 144. *'Sculpting' might not seem so appropriate a word here.* Unless, as I suggested half tongue-in-cheek in my very first publication in *Nature* (Dawkins, 1971), the observed (and rather alarming) phenomenon whereby large populations of brain cells die, every day, is non-random and a mechanism of memory. If I got it right, we could indeed speak of the brain as being sculpted, carved in a constructive way somewhat reminiscent of Darwinian natural selection.

P 145. *I think you can see the analogy between behaviour 'shaping' and Darwinian selection, the parallel that so appealed to Skinner and many others.* Skinner (1984), Pringle (1951).

P 145. *Darwin himself was an enthusiast of the pigeon fancy.* Darwin (1868). The Reverend Whitwell Elvin in 1859 told Darwin's publisher John Murray that instead of *Origin of Species* Darwin should have written a book entirely about pigeons. http.//friendsofdarwin.com/articles/whitwell-elwin/.

P 147. *A beaver would treat access to branches, stones, and mud suitable for dam-building as a reward.* My conjecture gains plausibility from a splendid film of a rescued young beaver building a dam out of Christmas decorations and toys. https.//laughingsquid.com/rescued-beaver-builds-indoor-dam/. I've also seen a German film of a beaver in a bare room building a non-existent dam out of thin air, a beautiful example of what ethologists call a 'vacuum activity'.

P 147. *The rats appeared to become addicted to lever-pressing.* Kringelbach & Berridge (2010).

P 149. *... quickly discovered that there was something erotic about the stimulation* (Frank, 2018).

P 153. *A respected ornithologist and philosopher named Charles Hartshorne suggested that we should regard birdsong as music.* Hartshorne (1958). See also Hall-Craggs (1969). And Roger Payne (1972) and Payne & McVay (1971) made a strong case for whale song being music, appreciated by whales and humans alike. Humpback whale songs are immensely long, loud, and complicated, as we know from Payne's beautiful recordings, used as inspiration by Judy Collins and other human musicians. Roger sadly died while I was in the final stages of writing this book. If whales could mourn (they probably can) and if they could read the many obituaries, I can hear in my imagination the Requiem they might compose for the man who did so much to save their kind from extinction at the hands of human greed. It is a consolation to me that Balliol College, Oxford, awarded him the 2007 Dawkins Prize for Animal Conservation and Welfare.

P 153. *The role of learning and genes in the development of birdsong has been intensively studied.* Thorpe (1961), Marler & Slabbekoorn (2004), Catchpole & Slater (2008), Kroodsma & Miller (1982).

P 153. *The 'Beau Geste' hypothesis of John Krebs.* Wren (1924), Krebs (1977).

P 154. *Virtuoso impressionists like lyre birds, mynahs, parrots, and starlings.* I don't know if this is what gave him the idea, but John Krebs himself had a pet starling in his room at Oxford, and it would disconcert his tutorial pupils by saying, in a pitch-perfect imitation of John's own voice, 'Yes … yes … yes …' every time there was a pause in the student's discourse. It also did a tantalising imitation of liquid being poured from a bottle, and it performed the opening bars of a Mozart aria. Starlings are adept at imitating telephone ringtones, and they are responsible for many a gardener rushing indoors to no purpose. Such incidents seem to add plausibility to the Beau Geste hypothesis. It may be significant that starlings are outliers among songbirds, with abnormally complex songs. Each male has about sixty to eighty 'motifs', many of them learned by imitation, woven into a highly complex song.

P 154. *Song sparrows … were reared by canaries in a soundproof room so that they never heard the song of a song sparrow.* Konishi & Nottebohm (1969).

P 155. *This [deafening] has been done, with both song sparrows and the related white-crowned sparrows.* Konishi & Nottebohm (1969).

P 156. *As a student I tried to read up that rat psychology literature.* It wasn't only rats. But it might as well have been. One of the journals, which I shall refrain from naming, always referred to the animals being studied as 'Ss' (short for 'subjects'). Somewhere near the beginning of the 'Methods' section of any paper, you'd see, 'Ss were rats …' or 'Ss were pigeons …' The point was that, practical convenience aside, it didn't matter which species you were studying. Ss were Ss. Regardless of species, they were empty vessels into which the laws of learning were poured.

P 156. *A pair of papers that I wrote jointly with John Krebs some years ago, about animal signals generally.* Dawkins & Krebs (1978), Krebs & Dawkins (1984).

P 156. *He sits and sings, and the female comes to him under her own power.* Dawkins & Krebs (1978). Merlin Sheldrake (2020) makes a similar point with respect to the fungus *Ophiocordyceps*, which invades the body of an ant (Hughes et al., 2012). It somehow drugs the ant, causing it to climb to the top of the nearest plant and clamp its jaws – the so-called 'death grip' – in a major vein of the plant. Feeding on the ant, the fungus sprouts a little mushroom from the ant's head, which discharges spores that rain down on any ants that might be passing below. In Sheldrake's words, 'The fungus doesn't have a twitchy, muscular animal body with a centralised nervous system or an ability to walk, bite or fly. So it commandeers one. It is a strategy that works so well that it has lost the ability to survive without it. For part of its life, *Ophiocordyceps* must wear an ant's body.'

P 157. *This concerns what we called 'mindreading'.* Krebs and Dawkins (1984).

P 158. *Ultimately because sperms are smaller and more numerous ('cheaper') than eggs, females need to be choosier than males.* Trivers (1972), Symons (1979), Low (2000), Miller (2000).

P 158. *I've been a bit terse in condensing two full-sized papers into four paragraphs.* Not to mention the chapters called 'Arms Races and Manipulation' and 'Action at a Distance' of my early book *The Extended Phenotype.*

P 160. *A week's exposure to a bow-cooing male reliably causes massive growth of a female's ovary.* Lehrman (1964). An interesting complication is that later experiments by Mei-Fang Cheng, in Lehrman's old department after his death, suggested that the male influences the female indirectly. His bow-cooing causes her to coo, and it's her own cooing that stimulates her ovaries to grow. I don't think this changes my argument, although it certainly complicates it (Cheng, 1986).

P 160. *Parallel work by Robert Hinde and Elizabeth Steel in Cambridge on nest-building behaviour in female canaries showed the same thing.* Hinde & Steel (1976).

P 162. *What Konrad Lorenz, one of the fathers of ethology, dubbed the 'Innate Schoolmarm'.* Lorenz (1966b).

P 162. *Joan Stevenson found that chaffinches preferred to settle on a perch attached to a switch which turned on chaffinch song.* Stevenson (1969).

P 162. *Braaten and Reynolds with hand-reared, naive zebra finches and using starling song for comparison.* Braaten & Reynolds (1999).

P 163. *Man is a nidicolous species.* This was stated in a paper by the distinguished Dutch ethologist and anthropologist Adriaan Kortlandt. Unfortunately, he mistyped it as 'nidiculous'. The editors of the journal (also Dutch, although their journal was published in English) were unable to reach Dr Kortlandt (studying chimpanzees in the depths of the African forest) so they had to take an editorial decision. 'Nidiculous' is a one-letter mutation away from both 'nidicolous' and 'ridiculous'. 'Ridiculous' is far more common in English than 'nidicolous' so they took refuge 'in the laws of probability' and printed Kortlandt's sentence as 'Man is a ridiculous species'. Of course, they knew perfectly well what he had meant. It was an act of humorous mischief at Kortlandt's expense. Probably not the last.

P 166. *They are labelled by the presence of adjacent nonsense sequences of DNA, which are palindromes.* A palindrome reads the same both forwards and backwards. Like the apocryphal epitaph for Napoleon, 'Able was I ere I saw Elba', or the first words spoken to Eve (herself a palindrome), 'Madam I'm Adam'.

P 166. *Scientists have discovered a way in which the bacterial skill can be borrowed for the human purpose of editing genomes.* Jennifer Doudna, one of the discoverers of CRISPR, has co-authored a nice personal account (Doudna & Sternberg, 2017).

P 168. *The details differ between Andean and Himalayan peoples.* Beall (2007).

P 169. *Octopuses and other cephalopod molluscs, who have perfected the art of dynamic cross-dressing to an astonishing extent.* Hanlon (2007).

P 175. *The brain in a vat ('Where am I?').* Daniel Dennett, reprinted in Hofstadter & Dennett (1981), imagined that his brain was removed and sustained in a vat, connected to his body by radio. Thought experiments of this kind convince me that some philosophers do worthwhile things. Especially if, like Dennett, they take the trouble to study science.

Chapter 8. The Immortal Gene

P 176. *It has become the working assumption of most field zoologists studying animal behaviour and behavioural ecology in the wild.* Ågren (2021) gives a scholarly and balanced account of the history of the gene's-eye view. Sterelny & Kitcher (1988) do the same from a philosophical point of view.

P 176. *Disagreement that is clearly stated deserves a clear reply.* Scientists who are wrong do us a useful service when they display their error in plain view, especially when that error is widely shared but not explicitly voiced. The Scottish zoologist VC Wynne-Edwards (1962) is justly celebrated for bringing out into the open an important error, which had hitherto been implicit, unclear, and lamentably widespread – for example, in the writings of the American ecologist WC Allee, and the Austrian ethologist Konrad Lorenz (1966a). The 'group selection' fallacy had been accepted without realising it by many predecessors. They implicitly assumed, and Wynne-Edwards explicitly stated, that natural selection chooses between *groups* of animals. Some assumed that individuals do whatever is best for the preservation of the species. Wynne-Edwards specifically proposed that individuals take steps to limit their personal reproduction because overpopulation is bad for the group. This was wrong, but I mention it only by analogy, to make the point that mistakes can be put to constructive use.

P 176–7. *This book will show ... natural causes.* Noble (2017), page x.

P 177. *Almost all scientists working in the fields known as Behavioural Ecology, Ethology, Sociobiology, and Evolutionary Psychology.* Alcock (1979), Barash (1982), Barkow, Cosmides & Tooby (1992), Bateson & Hinde (1982), Buss (2005), Chagnon & Irons (1979), Clutton-Brock et al. (1982), Daly & Wilson (1983), Gadagkar (1997), Grafen & Ridley (2006), Haig (2020), Halliday & Slater (1983), Hamilton (1996, 2001, 2005), Hinde (1982), King's College Sociobiology Group (1982), Krebs & Davies (1978, 1984, 1987, 1991), Low (2000), Manning and Stamp Dawkins (1998), McFarland (1985), Miller (2000), Symons (1979), Taborsky et al. (2021), Trivers (1985, 2011), Wilson (1975), Workman & Reader (2004).

P 178. *The doctrine of the relativity of ... in relation to other organs.* Singer (1931), p. 568.

P 178. *The principle of Biological Relativity is simply that there is no privileged level of causation in biology.* Noble (2017), p. 160.

P 181. *Before the invention of printing, biblical scriptures were painstakingly copied by scribes.* Tov (undated).

P 181. *Mutation is never biased towards improvement.* Some genes are more prone to mutate than others. Some regions of a genome are so-called hotspots where mutation rates are especially high. There are genes called 'mutator' genes whose phenotypic effect is to increase the mutation rate in other genes. The latter would be generally disfavoured by selection because most mutations are deleterious. A random change in a system which already works well is likely to make thing worse – 'If it ain't broke, don't fix it'. As the great evolutionist George C Williams argued, Darwinian selection will tend to push mutation rates towards zero, a result which is fortunately never achieved – fortunately, because evolution would cease if it were. A decrease in mutation rate is especially desirable in those regions of the genome housing genes that are more than usually important for the organism's welfare. Natural selection would be expected to work especially hard to 'protect' such regions from (random) mutation. We might therefore expect to find 'anti-hotspots' (mutational 'coldspots') in such vitally important regions of the genome. There is evidence of this from the plant *Arabidopsis*, the botanist's equivalent of the *Drosophila* fruit fly (Monroe et al., 2022).

P 181. *Mutation has no way to judge in which direction improvement lies.* In our public debate at Hay-on-Wye in 2022, Denis Noble reiterated his hope (Noble, 2017) that evidence might one day be found in favour of mutations guided in beneficial directions. I sceptically await any such evidence, very sceptically in the case of bodily adaptations in eucaryotic organisms reproducing sexually. In our 2022 debate at Hay-on-Wye, Noble stated, correctly, that Darwin, late in life, devised a Lamarckian theory of his own, 'pangenesis'. Noble eccentrically believes that this was the real Darwin, who would not have subscribed to neo-Darwinism, and would not have applauded the German zoologist August Weismann with his concept of an independent germline, flowing like a river through geological time, with a succession of mortal somas as side branches. I follow the majority of biologists and historians of science in regarding Darwin's 'gemmules' as an aberrant, misguided effort to rescue his theory from criticisms which, in the light of Mendelian genetics, we can now see to have been misguided. If only Darwin had read Mendel, a great burden would have been lifted from him, and pangenesis and gemmules would have gone out of the window. And I suspect that Darwin would have loved Weismann's vision of life. This is another respect in which I think Noble is flat wrong. I believe Darwin would be delighted by the neo-Darwinian revolution whereby Fisher and others joined Darwin's great work to Mendelian genetics.

P 181. *Most animals die in a matter of years if not months or weeks ... And their physical DNA molecules die with them. But the information in the DNA can last indefinitely.* Williams (1966a).

P 182. *With minor exceptions such as the Y-chromosome.* JBS Haldane wrote of his father (and therefore of himself) 'He was born with a historically labelled Y chromosome. That is to say, his ancestors in the putative direct male line since about A. D. 1250 are known. There are, I believe, about fifteen similarly labelled sets of Y chromosomes in Britain.' Theoretically, a man's surname is a historical label of the Y-chromosome, but the assumption of legitimate paternity in every generation since 1250 is an ambitious one, to say the least. The geneticist Bryan Sykes contacted a large sample of

Sykes men in Yorkshire and neighbouring areas and found that about 50 per cent of them had the same Y-chromosome as himself. He worked out that all were descended from an ancestral Sykes who lived in the thirteenth century (coincidentally around the same date as Haldane's known ancestor). He accounted for the 50 per cent who didn't share the same Y-chromosome by postulating illegitimacy (or equivalent non-paternity) at some point during the generations. The data enabled him to calculate the frequency of such events, and it came to between 1 and 2 per cent per generation. That's a low figure. But if there were, say, four generations of Haldanes per century, the chance that JBS's lineage ever since 1250 included at least one illegitimacy (or equivalent) would come to about 40 per cent. http.//cafamilies.org/sikes/bbc/surnames_prog1.html.

P 182. *They are sundered in every generation by the process of crossing over.* Sperms and eggs are made by a special kind of cell division, meiosis. The end result of meiosis is that each gamete has only one set of chromosomes – 23 in us – instead of the double set – a total of 46 in us – possessed by normal body cells. The 46 consist of 23 chromosomes inherited intact from the mother and 23 inherited intact from the father. The two sets of 23 remain independent of each other in all the cells of the main body. Meiosis brings the members of each pair together, lining them up opposite one another. And then a remarkable thing happens. They exchange substantial parts of their length. This is crossing-over. You can see that, because of crossing-over, whole chromosomes are not replicators. Small lengths of chromosome may replicate for many generations before crossing-over cuts them.

P 182. *It is no paradox that The Cooperative Gene would also have been appropriate.* So would *The Strategic Gene*, David Haig's (2002) suggestion.

P 185. *If there's life elsewhere in the universe it will be Darwinian life.* I put the argument in 'Universal Darwinism', my contribution to the Darwin centenary conference in Cambridge (Dawkins, 1983).

P 185. *It happens that on our planet the replicated information, the causal agent in the Darwinian process, is DNA.* It is arguable, though I won't press the point here, that cultural inheritance can mimic genetic inheritance, even affecting anatomical phenotype. Memes are another kind of replicator. In a rather odd sense, the circumcised phenotype does have a statistical tendency to pass down generations, because religion has that same trans-generational tendency and religion can cause circumcision. But the relevant question is whether *randomly* administered circumcision would replicate to the next generation. I suppose it might, if fathers want their sons to 'look like me'. Though mildly interesting as an analogy to genetic transmission, it's too trivial to affect the point I am trying to make, and too minor to outgrow an endnote.

P 186. *The metaphor of the bookkeeper has a dramatic appeal so seductive that it evidently seduced Gould himself.* Gould (1992) in his review of Helena Cronin's beautiful book, *The Ant and the Peacock* (1991).

P 186. *When the bookkeeper makes an entry in his ledger, the entry does not cause a subsequent monetary transaction. It is the other way around.* The exception that comically proves the rule: I once overheard the following lunchtime conversation between the Registrar (powerful senior administrative officer) of Oxford University and another Fellow of New College, Oxford. 'What happened at Hebdomadal Council yesterday?' 'I don't know. I haven't written the minutes yet.' That's not the way things are supposed to be done, it's not the way bookkeepers do their thing, and genes are not bookkeepers but active causal agents. Of course, he was joking.

P 188. *'Clade selection', a coining of George C Williams, fits the bill.* Williams (1992). Here's another possible example of clade selection. Most animals reproduce sexually, but asexual reproduction (females reproduce without any intervention from males) pops up sporadically (Maynard Smith, 1978). And 'sporadically' is what makes it interesting. If you draw a family tree of all life, and colour in the asexual branches, you'll notice that you are colouring the tips of twigs, not major branches. It looks as though, from time to time, asexual reproduction arises, then very soon goes extinct before a large clade has had time to evolve. As far as I know, there is only one example of a major clade consisting entirely of asexual females. This is the bdelloid class of the phylum Rotifera. I have heard John Maynard Smith say, in his inimitable way, that the bdelloid rotifers are a scandal and there should be a law against them.

P 188. *Well, you can't draw a [butcher's] map like that for domains of genes.* You can draw such a map for the domains of tactile sensory nerves on the skin, but that, though interesting, is another story. Genes aren't like that.

P 189. *A large sheet hangs from the ceiling, suspended from hooks by hundreds of strings.* My hanging sheet model looks superficially like Waddington's (e.g. 1977) 'epigenetic landscape'. But the two are doing very different things and should not be confused.

PP 191–2. *A long and rather complicated verbal definition [of inclusive fitness].* Hamilton (1964).

P 192. *In one of his papers, he wrote.* Hamilton (1972).

P 194. *But he substituted the word 'interactor' for my 'vehicle'.* Hull (1981).

P 194. *On the gene's-eye view, the very existence of vehicles should not be taken for granted but needs explaining in its own right.* See the chapter called 'Rediscovering the Organism' in Dawkins (1982).

P 194. *Replicators (on our planet, stretches of DNA).* And perhaps memes, and goodness knows what exotic replicator on other worlds, but this is not the place to go into that.

Chapter 9. Out Beyond the Body Wall

P 197. *Its astonishing building skills were unravelled by Michael Hansell.* Hansell (1968).

P 197. *These larvae are master masons.* Hansell (1984, 2007).

P 201. *Like the remarkable backwards-facing trombone of the dinosaur Parasaurolophus, which probably served as a resonator for its bellowings.* My six-year-old grandson's favourite dinosaur is *Parasaurolophus* and it was he who first drew my attention to it and pointed out its remarkable crest. Other Hadrosaur dinosaurs had a domed top to the head, which seems also to have been a resonator. There's an interesting convergence with a group of extinct wildebeest-like mammals, whose domed nasal cavity probably served the same purpose. O'Brien et al. (2016).

P 201. *Henry Bennet-Clark showed that the double horn concentrates the sound.* Bennet-Clark (1970).

P 145. *Behaviour genetics arouses a scepticism never suffered by anatomical genetics.* This may be partly because a certain stamp of confused mind thinks it is associated with racism.

P 203. *Penetrating genetic research by David Bentley, Ronald Hoy, and their colleagues in America.* Bentley & Hoy (1974).

P 206. *Something like it [the evolution of the mole cricket's double megaphone] must have happened through natural selection.* You could give natural selection a helping hand, I conjecture. You could render all female mole crickets in a large enclosure partially deaf, thereby upping the selection pressure on males to sing louder. Crickets keep their ears on their legs, and humans go a little deaf if we build up too much wax in our ears, so perhaps you could make the female crickets hard of hearing by painting their legs with a waxy film. I predict that if a long-term experiment were done along these lines, waxing female legs successively over enough generations, natural selection would come to favour male crickets that increase the amplification of their calling song by digging bigger megaphones. If such further amplification is possible, you might say, why don't males do it anyway, even without the females being waxed? The answer (and there's a general lesson here) probably lies in the ubiquitously important notion of economic compromise. Digging a larger megaphone would cost extra energy. The precise extent of any Darwinian adaptation is the result of a delicate balance between benefits and costs. Artificial deafening of females will shift the point of balance.

P 208. *Bright coloration in males is favoured, either through attracting females or through besting rival males.* Cronin (1991), Andersson (1994).

P 208. *Bower birds are a family of birds inhabiting the forests of New Guinea and Australia.* Gilliard (1969).

P 212. *The special issue of the journal [on The Extended Phenotype] came out in 2004.* Laland (2004), Turner (2004), Jablonka (2004), Dawkins (2004).

Chapter 10. The Backward Gene's-Eye View

P 215. *'Shouts all day at nothing'.* AE Housman, *Last Poems*, XL.

P 215. *My main authority – indeed today's world authority – is Professor Nick Davies.* Davies (2015).

P 216. *There is, of course, no suggestion that it knows what it's doing, or why it is doing it, no feelings of guilt or remorse (or triumph) in the act.* We cannot know what, or whether, they think or feel. We don't deny it (Griffin, 1976). Rather than assume that they don't, or do, ethologists temporarily ignore the question and concentrate on what we can observe and measure. Natural selection, too, sees only behaviour. If it selects between feelings, it can be only indirectly, via the behaviour that the feelings engender. The honeyguides are unrelated brood parasites from Africa and Asia, and they use a different, if anything more gruesome, murder method. They have sharp hooks on their beaks with which they slash and cut their nestmates to death. That's if they have any nestmates, after puncturing the host eggs with the same sharp hooks. By the way, nothing to do with their brood parasitism but too interesting to pass by, these birds are named for their remarkable habit of guiding humans to bee nests so that, when the humans have broken into the nest for honey, the birds can cash in and eat the wax and grubs. They have a special call which, in effect, means 'Follow me to honey'. I'm intrigued by this mutualism and wonder how far back in evolution it goes. Did these birds also guide our Australopithecine ancestors and their predecessors in Pliocene Africa? It might seem plausible, since early hominins would have been more adept than us at climbing trees, where bee nests naturally occur. On the other hand, one of the reasons honeyguides need human help in raiding bee nests is that humans calm the bees with smoke. There's no evidence that *Australopithecus* tamed fire. *Homo erectus* probably did, so conceivably the human/honeyguide partnership dates back a million years before *Homo sapiens*, even if no further. A million years would perhaps be time enough for natural selection to build the behaviour into the honeyguide's repertoire. Incidentally (Yong, 2011), the widespread belief that honeyguides also guide honey badgers (ratels) to bee nests seems to be based on no evidence.

P 216. *The hysterically unpopular downgrading of Pluto to sub-planet status.* Tyson (2014).

P 217. *I'm happy to quote no less an authority than Beethoven in support of my hearing it as a major third.* Composers are not unanimous. Mahler, in his first symphony, renders the descending interval as a perfect fourth. Handel's ('Cuckoo and Nightingale') organ concerto has a minor third. According to some descriptions, cuckoos start with a minor third in spring but stretch it to a major third in summer. Sure enough, Delius has a minor third in his 'On hearing the first cuckoo in spring' and I don't think he wrote a follow-up in summer. May we conclude that Beethoven's pastoral idyll was of summer? As the old rhyme has it, 'The cuckoo comes in April, / Sings the month of May. / Changes its tune / In the middle of June. / In July he flies away.'

P 219. *The actual size [of the cuckoo egg] is a compromise.* Another example of the ubiquity of compromise, and one close to home, concerns the human pelvis and baby's head. Our Australopithecine ancestors renounced the quadrupedal gait of all other primates. The selection pressure towards bipedality, whatever it was (See Kingdon, 2003), modified the pelvis in ways that favoured fast bipedal running, to replace the quadrupedal cantering that we see in other primates and well displayed by baboons. At the same time, bipedality freed the hands to shape tools and carry and manipulate objects. This created conditions favourable to the evolution of intelligence and brain enlargement. But large brains in babies made childbirth difficult, which put pressure on the female pelvis to get bigger (and put pressure on babies to be born earlier and more helpless). The best pelvis to give birth is not the best pelvis for running fast. Inevitably, the female pelvis is a compromise between two opposing selection pressures, too large for optimal locomotion, too small for optimal childbirth. Helen Joyce (2021) uses this evolutionary argument in buttressing the already strong case for separate sporting events for biological females.

P 222. *With the exception of W-chromosome genes.* And mitochondrial genes, but that's probably irrelevant.

P 224. *The aerial swerving and dodging chases of Spitfires and Messerschmitts.* A more than usually vivid parallel is the nocturnal arms race between noctuid moths and the bats that hunt them by radar. The bats' 'radar' is actually sonar, of course, and the moths have evolved ears that are tuned specifically to the ultrasound frequencies that bats use. When a moth hears anything at all, it can be assumed to be a bat, and it triggers a sequence of diving, spiralling, dodging, twisting manoeuvres reminiscent of those employed by human pilots in aerial combat. (Roeder & Treat, 1961).

P 225. *I can't help wondering whether Chaucer was using 'heysugge' to mean LBB rather than specifically Prunella modularis.* AS Kline's translation into modern English renders 'heysugge' as 'sparrow'. Perhaps because 'hedge sparrow' would have mucked up the scansion.

P 226. *Sex-linked genes are those that are actually carried on the sex-chromosomes themselves.* Our Y-chromosome is small, and there are very few known examples of sex-linked Y genes. Hairy ears used to be the familiar textbook example, but even this has been called in question. No matter if there are no male sex-linked characteristics at all, there are huge numbers of male sex-limited characteristics, including not just obvious things like the penis, but statistical influences on body size, muscle development, running speed, swimming speed, tennis serve strength, and many other characteristics.

P 234. *Enough to compete with siblings, but not so much as to overtax itself or attract predators.* The Israeli zoologist Amotz Zahavi (1997) proposed the intriguing theory that attracting predators is actually the name of the game. The nestling is blackmailing his parents to feed him in order to shut him up lest his loud squawks exactly do attract predators! His squawks, translated into English, are saying, 'Cat, cat, come and get me! I am here and I don't care who knows it until my parents feed me.' Initially sceptical, I now like the theory, but it makes no difference to my discussion of cuckoos.

P 235. *Niko Tinbergen reported that oystercatchers, offered a choice, will preferentially attempt to incubate a dummy egg eight times the volume of their own egg.* Tinbergen (1951).

P 237. *In addition to its yellow gape, it has a pair of dummy gapes: a patch of bare skin on each wing, the same yellow colour as the real gape.* Tanaka et al. (2005).

P 238. *There's a similar story for another brood parasite, the whistling hawk cuckoo.* Li et al. (2010).

Chapter 11. More Glances in the Rear-View Mirror

P 241. *Pushes this way of thinking to the limit.* Haig (1993, 2002, 2020).

P 241. *Ways in which genes within an individual can come into conflict with one another.* The whole topic of *Genes in Conflict* is comprehensively covered in the book of that name by Burt and Trivers (2006). Robert Trivers is one of the seminal thinkers who inspired a whole generation of evolutionary biologists including David Haig (and me).

P 241. *Two-thirds of your ancestral history has been in female bodies, one-third in male bodies.* Shaffner (2004).

P 242. *Only eight [elephant seal] males inseminated an astonishing 348 females.* Le Boeuf (1974), Le Boeuf & Reiter (1988).

P 245. *There are other complications.* Charnov (1982).

P 246. *She only has one pup in a year, so she'll maximise her reproductive success by surviving herself.* Readers (and there will be many, unlike me) who think first and foremost of the human species will spot an apparent anomaly. Men can go on reproducing into more advanced old age than women, whose reproductive life is cut short in midlife by the menopause. It is doubtful whether, in a state of nature, a significant number of men lived long enough to avail themselves of this apparent advantage. And the menopause probably has its own advantage, occurring at a time when a woman can benefit her genes more by caring for grandchildren than for her own children.

P 246. *The 'sneaky male' strategy.* This is the polite version of the technical term. John Krebs and I were momentarily proud of introducing the less polite version into the scientific literature, but it was perhaps too easy since one of us was editor of the publication concerned. In an effort to atone for thus shooting a sitting bird, I'll stick to the polite version here.

P 248. *I perhaps went too far when in 1989 I published a speculation about naked mole rats.* Dawkins (1989).

P 249. *Even I am not foolhardy enough to predict rodents with wings.* But if bats hadn't got there first, there probably would be rodents with wings.

P 249. *Is it conceivable … might turn out to be the lost caste of the naked mole rat.* Dawkins (1989).

P 249. *Change utterly, and a terrible beauty is born.* WB Yeats, 'Easter, 1916'. 'Oh, in a moment' is also Yeats, but a different poem.

P 250. *The best way to visualise it is not on paper but zooming around a computer screen.* I strongly recommend having a zoom around the brilliant Zoompast program written by James Rosindell with Yan Wong.

P 250. *The royal haemophilia gene can be traced back to a particular individual ancestor, Queen Victoria, one of whose two X-chromosomes bore the gene.* Bodmer & McKie (1994).

P 251. *Irene married her first cousin Henry, a common practice among royals, and generally not a good idea because of inbreeding depression.* All of us have a few lethal or sub-lethal recessive genes. But these are rare so, if we mate at random, a child is unlikely to suffer from a double dose. If you marry your sibling, however, there is a 25 per cent chance of a child inheriting a double dose, and that applies to each one of your lethal or sub-lethal genes. If you marry your first cousin, the chances are one in sixteen for each deleterious gene, still high enough to counsel against the practice. In Pakistan, nearly 50 per cent of marriages are between first cousins. The custom persists among British people of Pakistani origin, and their infant mortality is double the national average. In addition to lethal recessive genes, which actually kill you, the more common sub-lethal recessives have debilitating effects both physical and mental. Charles Darwin was aware of inbreeding depression, although its cause was not understood in his time, and he fretted about his own possible unwisdom in marrying his first cousin Emma Wedgwood. Laboratory white rats are the product of many generations of brother–sister matings, and they are remarkably free of inbreeding depression. This is no paradox. The lethal and sub-lethal recessives have disappeared because of strong natural selection against them over previous generations. Of course, it's obvious when you think about it, a rare, authentic case of the exception proving the rule!

P 252. *For a rather odd reason, I was one of the earliest people in Britain to have their entire genome … sequenced.* In 2012 I was the presenter of a three-part TV documentary on Britain's Channel 4 called *Sex, Death and the Meaning of Life.* For one of the three episodes, the original plan was that my entire genome should be sequenced and the data disc buried in the Dawkins family vault in Chipping Norton church for 1,000 years. On screen, I was to imagine that it would be dug up a millennium hence, and a clone of me would be born. I was to ruminate on screen. What advice I would give to my young twin (for twin is what he would be)? 'Don't make the same mistakes I made! Make better use of our shared genome than I did.' It was also an opportunity to demystify the notion of a clone, and meditate on the question of personal identity. My clone would not be me, he would have his own personal identity just as monozygotic twins do. I would interview present-day identical twins to illustrate this. The young Richard of the future would grow up in a radically different world and he'd be able to tell me of all the amazing changes that had happened in the world during the thousand years that would intervene between us. And he, in imagination, could look back on the amazingly primitive world in which old Richard had lived, the changed mores, customs, technology, and language. As it turned out, the documentary took a different turn, but by then the production company had paid for my genome to be sequenced and the data disc was presented to me as a welcome gift.

P 254. *When Dr Wong did this with my genome, he found that a large majority coalesced somewhere around 60,000 years ago, say 50,000 to 70,000.* Dawkins & Wong (2016), p. 68.

P 255. *One was able to use the genome of the other to make a quantitative estimate of prehistoric demography affecting not just one individual but millions.* Dr Wong adds this note: 'It might at first glance seem surprising that your Jean and John chromosomes reveal general features of human history, rather than history specific to your and your close family. You can think of that as a feature of the deep age of Jean and John's genetic common ancestry, which is measured in thousands, tens of thousands, or even hundreds of thousands or millions of years. By the time you go back even 1,000 years, you have so many great-great-great … grandparents that they can be thought of as being a random sample of most Europeans at that time. As you go back even further, your ancestors become essentially like a random sample of all non-Africans (or at least, those that have not been reproductively isolated for long periods).'

P 256. *Fifteen per cent of the human genome is closer to the gorilla genome than it is to chimpanzee, and 15 per cent of the chimpanzee genome is closer to the gorilla than human.* Scally (2012).

P 257. *It's even theoretically possible (though vanishingly improbable) that John received all his genes from two of his grandparents, and none from the other two.* Svante Pääbo amused himself by calculating the probability of a Neanderthal walking into his office. Europeans typically contain about 2 per cent Neanderthal genes, but it's a different 2 per cent in each of us. Theoretically, all the different 2 per cents could chance to come together in a single individual. It's not going to happen!

P 257. *You are quite probably descended from William the Conqueror.* If you go back in time sufficiently far and locate anybody, that person is the ancestor of either everybody alive today, or of nobody alive today. There are no half measures. The rationale for this surprising conclusion is given in *The Ancestor's Tale*, 'Rendezvous 0'. The only question here is whether William the Conqueror is sufficiently far in the past.

Chapter 12. Good Companions, Bad Companions

P 260. *The genes within any one gene pool are travelling bands of good companions.* Yanai & Lercher (2015).

PP 261–2. *Every animal ever born throughout evolutionary history would have been classified in the same species as its parents.* I spelled out the argument in Dawkins (2011). There are exceptions, especially in plants, but I shall not deal with them here.

P 262. *The split originates accidentally, imposed perhaps by a geographic barrier such as a mountain range or a river or stretch of sea.* The initial barrier is not always geographic, especially among insects, which often split in what is called 'sympatric' speciation. Also, sympatric speciation seems to be important in the adaptive radiations of lake fish (Schluter & McPhail, 1992), including the Cichlid fishes of Africa's great lakes, which I mentioned in another connection in Chapter 6.

P 267. *My friend and former student Mark Ridley wrote a fine book with that very title.* In the American edition (2001). As is so often the way, it is different from the original British edition, which is *Mendel's Demon* (2000). Mark is not to be confused (though they often are) with Matt Ridley (no relation), also a close friend and author of fine books. A journal editor once managed the feat of getting them, unknowingly, to review each other's books in the same issue. Both were complimentary, and Mark ended his review by saying that Matt's book would be 'an excellent addition to our joint CV'.

P 270. *EB Ford, the eccentrically fastidious aesthete* (Ford, 1975). And, it must be admitted, snob. When referring to an easy piece of mathematics, where others might say 'as we learned in kindergarten', Ford's utterly characteristic form of words was, 'as we learned at our nursery-maid's knee'.

P 270. *A lifelong authority on butterflies and moths.* It was Ford's popular book on butterflies (1945) that introduced genetics to the young Bill Hamilton. Alan Grafen (2005), in his biographical addendum to Hamilton's three-volume memoir, remarked that 'To have inspired this one young biologist would by itself justify Ford's efforts in writing *Butterflies*.'

P 272. *Named after the entomologist and artist John Curtis (1791–1862).*

Curtis's own painting, in his eight-volume treatise on British insects (1832), is labelled *Triphaena consequa*. He found the specimen on the Isle of Bute, off western Scotland. The rules of taxonomy have the consequence that the official Latin name of a species can change, if an earlier naming is discovered. What was *Triphaena consequa* in 1832 could have another name today. A modern commentary on Curtis's book gives *Noctua comes ab. curtisii* as a synonym for the species that Curtis called *Triphaena consequa*. *Noctua* was the earlier name for the genus, given by Linnaeus, the father of taxonomy, and now restored as the official name. '*Ab.*' is short for 'aberrant form', and you could certainly describe the dark-winged morph as aberrant when compared with the common paler form. I conclude that the dark moth in the painting, which Curtis labelled *Triphaena consequa*, is none other than what Ford knew as the 'curtisii' morph of *Triphaena (Noctua) comes*.

P 274. *The uneasy pandemonium of genes within the genome, sometimes cooperating, sometimes disputing, is captured in Egbert Leigh's 'Parliament of Genes'.* (Leigh, 1971).

P 275. *The tiny roundworm Caenorhabditis elegans.* Molecular geneticists often like to refer to *Caenorhabditis elegans* as 'the' nematode or even 'the' worm, as though there were no other. It is actually only one of more than 50,000 species of worm and 30,000 species of nematode. Ralph Buchsbaum's textbook of invertebrate zoology (1971) quotes the following memorable image. 'If all the matter in the universe except the nematodes were swept away, our world would still be dimly recognizable, and if, as disembodied spirits, we could then investigate it, we should find its mountains, hills, vales, rivers, lakes, and oceans represented by a film of nematodes. The location of towns would be decipherable since, for every massing of human beings, there would be a corresponding massing of certain nematodes. Trees would still stand in ghostly rows representing our streets and highways. The location of the various plants and animals would still be decipherable, and, had we sufficient knowledge, in many cases even their species could be determined by an examination of their erstwhile nematode parasites.' Quite apart from nematodes, moreover, there are at least four other phyla of 'worms',

P 277. *That is to say they are epigenetically different while being genetically the same.* The Edinburgh geneticist, embryologist, and theoretical biologist CH Waddington originated the term 'epigenetics' in 1942. It is the study of the differential expression of genes in embryology – how genes are, or are not, switched on in different cells – as opposed to genetics, which is the study of the presence or absence of genes themselves in successive generations. The waters have been tiresomely muddied recently by grandiloquent popular-science authors who use 'epigenetics' preferentially, even exclusively, for those rare, and in my opinion insignificant, special cases where epigenetic switch-ons or switch-offs of genes carry over to the next generation. Waddington derived 'epigenetics' from 'epigenesis', a historical school of embryology to which all modern embryologists subscribe, as opposed to preformationism (the now dead theory according to which the egg, or in another version the sperm, contained a miniature embryo all ready to swell into the full body). Given that, as we now know, the cells in different tissues contain the same genes yet are so different from each other, there would seem to be no logical alternative to differential switching on or off of genes, i.e. epigenetics in Waddington's sense.

P 282. Generations of Oxford zoologists will never forget EB Ford's lecture on the strange Boundary Phenomenon in the butterfly *Maniola jurtina*. Ford and his colleagues found an abrupt discontinuity between two stable polymorphisms on either side of a line crossing south-west England (Ford, 1975). In my terms these would be two alternative sets of 'good companion' genes, similar to the Barra/Orkney separation except that in this case there was no geographical explanation for the mysterious dividing line. Indeed, the line shifted from year to year. While tracking it on foot one year, they came upon a hedge which seemed to mark the boundary. My memory can recreate Professor Ford's fluting, ultra-precise diction: 'At this point it was clear that matters were becoming critical. We therefore sat down by the hedge and ate our sandwiches.' I strongly suspect that the hedge in the photograph is the very hedge in question. Ford would surely have lost no opportunity to take his hero RA Fisher to see it for himself. If I am right, it is a historic photograph.

Chapter 13. Shared Exit to the Future

P 283. *In the words of Jake Robinson's title.* Robinson (2023).

P 283. *It is now established beyond doubt that they [mitochondria] originated from free-living bacteria.* Margulis (1998).

P 285. *The key to companionable benevolence … or its reverse, is the route by which a gene travels from a present body into a body of the next generation.* Fine (1975), Ewald (1987, 1994).

P 285. *In a large survey of the distribution of IRBPs in more than 900 species.* Kalluraya et al. (2023). The work was done in the laboratory of Matthew Daugherty.

P 285. *Amphioxus, a small, primitive creature.* 'Primitive' has a precise meaning in biology. It doesn't mean ancestral, and it is not derogatory. It means resembling ancestors. Amphioxus (*Branchiostoma*) is a modern, living animal, so it obviously cannot be ancestral to equally modern vertebrates. But it has changed less than modern vertebrates since the time of their shared ancestor. That defines it as primitive.

P 286. *A family tree of IRBP-like molecules.* Kalluraya et al. (2023).

P 287. *The parasitology literature is filled with macabre stories of parasites manipulating host behaviour.* Hughes et al. (2012). Dawkins (1990).

P 287. *Belonging to the phylum Nematomorpha.* Not nematodes or nemertines. These are three different phyla, easily confused by a similarity of names derived from *nema*, Greek for thread.

P 287. *Toxoplasma then needs the infected rat to be eaten by a cat, to complete its life cycle.* You might ask why parasites, especially worms, tend to have such complicated life cycles, going via intermediate hosts, sometimes as many as five intermediate host stages, on their way to the so-called definitive host. I think it's similar to the reason plants borrow animals to transport seeds or pollen. Spreading through the air in droplets from a cough or sneeze is all very well for a bacterium or virus, but worms need a larger vector such as an animal. Cats don't eat cats, but they do eat rats, and rats are conveniently mobile vectors for worms.

P 288. *The worm invades the eye stalk of the snail, which swells grotesquely, and seems to pulsate vividly along its length.* Simon (2014).

P 289. *Darwin . . . misdiagnosed the affinities of Sacculina.* Deutsch (2009).

P 290. *The nauplius larva is followed by the cyprid larva, and both are unmistakeably crustacean.* Calman (1911).

P 290. *Sacculina's genome has now been sequenced.* Blaxter et al. (2023).

P 292. *Parasitised ants also live longer, up to three times longer, than normal ants.* LePage (2023).

P 292. *It 'wants' the snail to shift its priorities towards individual survival. Hence, I suggest, the thickened shell.* Dawkins (1982, pp. 210–212).

P 294. *The familiar fact that viruses can help us as well as harm us.* Pride (2020). Each one of us is home to about 380 trillion viruses, many of them being bacteriophages (phages) that benefit us by preying on bacteria. A promising line of medical research is looking to phages that might save us from antibiotic-resistant bacteria.

P 295. *It has been estimated that about 8 per cent of the human genome actually consists of viral genes that have, over the millions of years, become incorporated.* Arnold (2020).

P 295. *It has been suggested that the evolutionary origin of the mammalian placenta was the result of a beneficial cooperation with an 'endogenous' retrovirus.* Haig (2012), Villafrreal (2016), Chuong (2018).

P 295. *LP Villareal … has gone so far as to say …* Villarreal (2016).

P 295. *'Duplicate me by the roundabout route of building an elephant first'.* The quotation is from Dawkins (1996).

P 296. *We live in a dancing matrix of viruses.* Thomas (1974).

P 296. *Barbara McClintock won a Nobel prize for her discovery of these 'mobile genetic elements'.* Pray & Zhaurova (2008).

P 296. *Some 44 percent of the human genome consists of such jumping genes or 'transposons'.* Mills et al. (2007).

Acknowledgements

The following read various drafts, in whole or in part, and offered helpful suggestions: John Krebs, Nick Davies, Jane Sefc, Michael Rodgers, Yan Wong, Joan Stevenson-Hinde, David Haig, and the extraordinarily meticulous Karen Owens.

Advice on particular points was kindly provided by Henry Bennet-Clark, Michael Hansell, Paula Kirby, Claire Spottiswoode, Ben Sandkam, Stephen Simpson, Peter Slater, Michael Ward, Latha Menon, Victor Flynn, Michael Kettlewell, Steven Balbus, Nicholas Kettlewell, and Ron Hoy. I am especially grateful to Edward Holmes for his advice and encouraging comments on Chapter 13.

In addition to the publishers' acknowledgement of photograph sources, the following went out of their way to be helpful in providing pictures: Keita Tanaka, Kathryn Marguy, Danielle Czerkaszyn, Michael Sweet, Anil Kumar Verma, Hussein Latif, Yathin Krishnappa, Tim Coulson, and Christopher Barnhart.

Author and artist have been well served by Head of Zeus, following initial encouagement from Anthony Cheetham. In addition to the many working behind the scenes, we should single out Neil Belton, Clémence Jacquinet and Jessie Price. For the US publication, it has been a pleasure to work with Jean Thomson Black and her colleagues.

Bibliography

Adams, D (1980) *The Restaurant at the End of the Universe*. Picador, London.

Ågren, JA (2021) *The Gene's-Eye View of Evolution*. Oxford University Press, Oxford.

Aktipis, A (2020) *The Cheating Cell – how evolution helps us understand and treat cancer*. Princeton University Press, Princeton, NJ.

Alcock, J (1979) *Animal Behavior*. Sinauer, Sunderland, MA.

Andersson M (1994) *Sexual Selection*. Princeton University Press, Princeton, NJ.

Arnold, C (2020) The non-human living inside of you. *Nautilus*. Coldspring Harbor Laboratory, NY.

Balbus, SA (2014) Dynamical, biological and anthropic consequences of equal lunar and solar angular radii. *Proc. Roy. Soc. A*, 470.

Barash, DP (1982) *Sociobiology and Behavior*. Hodder & Stoughton, London.

Barkow, JH, Cosmides, L & Tooby, J (1992) *The Adapted Mind*. Oxford University Press, New York.

Barlow, GW (2000) *The Cichlid Fishes: nature's grand experiment in evolution*. Perseus, New York.

Barlow, HB (1961) Possible principles underlying the transformations of sensory messages. In WA Rosenblish (ed.), *Sensory Communication*. MIT Press, Cambridge, MA.

Barlow, HB (1963) The coding of sensory messages. In WH Thorpe & OL Zangwill (eds), *Current Problems in Animal Behaviour*. Cambridge University Press, Cambridge.

Barnhart, MC et al. (2008) Adaptations to host infection and larval parasitism in Unionoidea – *J.N. Am. Benthol. Soc.*, 27, 370–394.

Bateson, PPG & Hinde, RA (1982) *Current Problems in Sociobiology*. Cambridge University Press, Cambridge.

Beall, CM (2007) Two routes to functional adaptation: Tibetan and Andean high-altitude natives. *Proceedings of the National Academy of Sciences*, 104 (suppl. 1), 8655–8660.

Bennet-Clark, HC (1970) The mechanism and efficiency of sound production in mole crickets. *Journal of Experimental Biology*, 52, 619–652.

Bentley, D and Hoy, R (1974) The neurobiology of cricket song. *Scientific American*, 231, 34–44.

Blaxter, M et al. (2023) The genome sequence of the crab hacker barnacle, *Sacculina carcini* (Thompson, 1836). *Wellcome Open Research*, 8, 91.

Bodmer, WF & McKie, R (1994) *The Book of Man*. Little Brown, London.

Boonman, A et al. (2014) Nonecholocating fruit bats produce biosonar clicks with their wings. *Current Biology*, 24, 2962–2967.

Braaten, RF & Reynolds, K (1999) Auditory preference for conspecific song in isolation-reared zebra finches. *Animal Behaviour*, 58, 105–111.

Brenner, S (1974) The genetics of *Caenorhabditis elegans*. *Genetics*, 77, 71–94.

Brenner, S (2002) Nature's gift to science. *Nobel Lecture*, 8 Dec., reprinted (2003) in *ChemBioChem* 4, 683–687.

Brunhoff , J de (1935) *Babar's Travels*. Methuen, London.

Buchsbaum, R. (1971) *Animals Without Backbones, Volume 1*. Pelican, London.

Burt, A & Trivers, RL (2006) *Genes in Conflict*. Harvard University Press, Cambridge, MA.

Buss, DM (ed., 2005) *The Handbook of Evolutionary Psychology*. Wiley, New Jersey.

Cain, AJ (1989) The perfection of animals. *Biological Journal of the Linnean Society*, 36, 3–29. Reprinted from JD Carthy & CL Duddington (eds) (1966), *Viewpoints in Biology*, 4. Butterworth, Oxford.

Caldwell, RL & Dingle, H (1976) Stomatopods. *Scientific American*, Jan., 80–89.

Callaway, E (2015) Beloved *Brontosaurus* makes a comeback. *Nature Communications*, 7 April.

Calman, WT (1911) *Life of Crustacea*. Macmillan, New York.

Catchpole, CK & Slater, PJB (2008) *Bird Song*. Cambridge University Press, Cambridge.

Cavalli-Sforza, LL (2000) *Genes, Peoples and Languages*. Allen Lane, London.

Cavalli-Sforza, LL & Feldman, MW (1981) *Cultural Transmission and Evolution*. Princeton University Press, Princeton, NJ.

Chagnon, NA & Irons, W (eds, 1979) *Evolutionary Biology and Human Social Behavior: an anthropological perspective*. Duxbury Press, North Scituate, MA.

Charnov, EL (1982) *The Theory of Sex Allocation*. Princeton University Press, Princeton, NJ.

Chaucer, G (1382) *The Parlement of Foules*. Librarius.

Cheng, M-F (1986) Female cooing promotes ovarian development in Ring Doves. *Physiology and Behavior*, 37, 371–374.

Chun, Li (2020) Amazing reptile fossils from the marine Triassic of China. *Bulletin of the Chinese Academy of Sciences*, 24, 80–82.

Chuong, EB (2018) The placenta goes viral. Retroviruses control gene expression in pregnancy. *PLOS Biol.*, 16, October.

Clutton-Brock, TH et al. (1982) *Red Deer: behavior and ecology of two sexes*. Chicago University Press, Chicago.

Conway Morris, S (2003) *Life's Solution: inevitable humans in a lonely universe*. Cambridge University Press, Cambridge.

Conway Morris, S (2015) *The Runes of Evolution*. Templeton Press, Pennsylvania.

Cott, HB (1940) *Adaptive Coloration in Animals*. Methuen, London.

Coyne, JA (2009) *Why Evolution Is True*. Oxford University Press, Oxford.

Craik, KJW (1943) *The Nature of Explanation*. Cambridge University Press, Cambridge.

Crew, B (2014) Caterpillar an expert in mimicry. *Australian Geographic*, 17 April.

Cronin, H (1991) *The Ant and the Peacock*. Cambridge University Press, Cambridge.

Curtis, J (1832) British Entomology. J Pigott, London.

Daly, M & Wilson, M (1983) *Sex, Evolution and Behavior.* Willard Grant, Boston.

Darwin, C (1859) *On the Origin of Species.* Murray, London.

Darwin, C (1868) *The Variation of Animals and Plants under Domestication.* John Murray, London.

Darwin, C (1871) *The Descent of Man.* Appleton, New York.

Davies, NB (2015) *Cuckoo: cheating by nature.* Bloomsbury, London.

Dawkins, R (1971) Selective neurone death as a possible memory mechanism. *Nature*, 229, 118–119.

Dawkins, R (1976, 1989) *The Selfish Gene.* Oxford University Press, Oxford.

Dawkins, R (1982) *The Extended Phenotype.* Oxford University Press, Oxford.

Dawkins, R (1983) Universal Darwinism. In DS Bendall (ed.), *Evolution from Molecules to Man.* Cambridge University Press, Cambridge.

Dawkins, R (1988) The evolution of evolvability. In C Langton (ed.), *Artificial Life* Addison Wesley, Boston.

Dawkins, R (1990) Parasites, desiderata lists, and the paradox of the organism. In AE Keymer and AF Read (eds), *The Evolutionary Biology of Parasitism. Supplement to Parasitology*, 100, S63–S73.

Dawkins, R (1996) *Climbing Mount Improbable.* Viking, London.

Dawkins, R. (2004) Extended phenotype – but not too extended. *Biology & Philosophy*, 19, 377–396.

Dawkins, R (2009) *The Greatest Show on Earth.* Free Press, London.

Dawkins, R (2011) *The Magic of Reality.* Transworld, London.

Dawkins, R (2013) *An Appetite for Wonder.* Bantam, London.

Dawkins, R & Krebs, JR (1978) Animal signals: information or manipulation. In JR Krebs & NB Davies (eds), *Behavioural Ecology*, 282–309.

Dawkins, R & Krebs, JR (1979) Arms races between and within species. *Proc. Roy. Soc. Lond. B*, 205, 489–511.

Dawkins, R & Wong, Y. (2016) *The Ancestor's Tale: a pilgrimage to the dawn of life.* Second Edition, Weidenfeld & Nicolson, London.

Dennett, D (1991) *Consciousness Explained.* Little Brown, Boston.

Deutsch, J (2009) Darwin and the Cirripedes: insights and dreadful blunders. *Integrative Zoology*, 4, 316–322.

Deutsch J (2010) Darwin and barnacles. *Comptes Rendus Biologies*, 333, 99–106.

Deutsch, JS & Mouchel-Vielh, E (2003) Hox genes and the crustacean body plan. *BioEssays*, 25, 878–887.

Diamond, J & Bond, AB (2013) *Concealing Coloration in Animals.* Harvard University Press, Cambridge, MA.

Doolittle, WF & Sapienza, C (1980) Selfish genes, the phenotype paradigm and genome evolution. *Nature*, 284, 601–603.

Doudna, JA & Sternberg, SH (2017) *A Crack in Creation: gene editing and the unthinkable power to control evolution.* Houghton Mifflin Harcourt, Boston.

Ewald, PW (1987) Transmission modes and evolution of the parasitism–mutualism continuum. *Annals of the New York Academy of Sciences*, 503, 295–306.

Ewald, PW (1994) *Evolution of Infectious Disease.* Oxford University Press, New York.

Feigin, CY et al. (2023) Convergent deployment of ancestral functions during the evolution of mammalian flight membranes. *Science Advances*, 9.

Fine, PEF (1975) Vectors and vertical transmission: an epidemiological perspective. *Annals of the New York Academy of Sciences*, 266, 173–194.

Fisher, RA (1930, 1958) *The Genetical Theory of Natural Selection*. Dover, New York.

Ford, EB (1945) *Butterflies*. Collins, London.

Ford, EB (1975) *Ecological Genetics*. Chapman and Hall, London.

Fowler, HW (1968) *Modern English Usage*. Oxford University Press, Oxford.

Framond, L de et al. (2022) The broken-wing display across birds and the conditions for its evolution. *Proceedings of the Royal Society B*, 289.

Frank, L (2018) Can electrically stimulating your brain make you too happy? *Atlantic*, 21 March.

Frisch, K von (1950) *Bees – their vision, chemical senses, and language*. Cornell University Press, Ithaca, NY.

Gadagkar, R (1997) *Survival Strategies*. Harvard University Press, Cambridge, MA.

Gallagher, P (2020) Be still my heart: dolphins can detect babies in the womb. *Evie Magazine*, 1 Oct.

Garvey, KK (2014) Can a bee unscrew the sting? *Bug Squad*, 24 Feb.

Gilliard, ET (1969) *Birds of Paradise and Bower Birds*. Weidenfeld & Nicolson, London.

Gissler, CF (1884) The crab parasite, *Sacculina*. *American Naturalist*, 18, 225–229.

Glenner, H et al. (2008) Induced metamorphosis in crustacean y-larvae: towards a solution to a 100-year-old riddle. *BMC Biology*, 6, 21.

Gould, SJ (1991) *Bully for Brontosaurus*. Hutchinson, London.

Gould, SJ (1992) The confusion over evolution. *New York Review of Books*, 39 (19), 47–54.

Grafen, A (2005) William Donald Hamilton. In Mark Ridley (ed.), *Last Words*. Volume 3 of WD Hamilton (2005), *Narrow Roads of Gene Land*. Oxford University Press, Oxford.

Grafen, A & Ridley, Mark (2006) *Richard Dawkins: how a scientist changed the way we think*. Oxford University Press, Oxford.

Grant, P & Grant, R (2014) *Forty Years of Evolution*. Princeton University Press, Princeton, NJ.

Gregory, R (1981) *Mind in Science*. Weidenfeld & Nicolson, London.

Gregory, R (1998) *Eye and Brain*. Oxford University Press, Oxford.

Griffin, DR (1959) *Echoes of Bats and Men*. Anchor, New York.

Griffin, DR (1976) *The Question of Animal Awareness*. Rockefeller University Press, New York.

Haeckel, E (2017) *The Art and Science of Ernst Haeckel*. Taschen, Cologne.

Haig, D (1993) Genetic conflicts in human pregnancy. *Quarterly Review of Biology*, 68, 495–532.

Haig, D (2002) *Genomic Imprinting and Kinship*. Rutgers University Press, New Brunswick, NJ.

Haig, D (2012) Retroviruses and the placenta. *Current Biology*, 22, R609–R613.

Haig, D (2020) *From Darwin to Derrida: selfish genes, social selves, and the meanings of life*. MIT Press, Cambridge, MA.

Haldane, JBS (1932) *The Causes of Evolution*. Longmans, Green, London.

Haldane, JBS (1940) Man as a sea beast. In *Possible Worlds*. Evergreen Books, London.

Halliday, TR & Slater, PJB (eds, 1983) *Animal Behaviour*. Blackwell Scientific Publications, Oxford.

Hall-Craggs, J (1969) The aesthetic content of bird song. In RA Hinde (ed.), *Bird Vocalizations*. Cambridge University Press, Cambridge.

Hamilton, WD (1964) The genetical evolution of social behaviour, I. *Journal of Theoretical Biology*, 7, 1–16.

Hamilton, WD (1972) Altruism and related phenomena, mainly in social insects. *Annual Review of Ecology and Systematics*, 3, 193–232.

Hamilton, WD (1996, 2001, 2005) *Narrow Roads of Gene Land*. Oxford University Press, Oxford. Three volumes.

Hamilton, WD & May, RM (1977) Dispersal in stable habitats. *Nature*, 269, 578–581.

Hanlon, R (2007) Cephalopod dynamic camouflage. *Current Biology*, 17, R400–R404.

Hansell, MH (1968) The house building behaviour of the caddis-fly larva *Silo pallipes* fabricius: I. The structure of the house and method of house extension. *Animal Behaviour*, 16, 558–561.

Hansell, MH (1984) *Animal Architecture and Building Behaviour*. Longman, London.

Hansell, MH (2007) *Built by Animals: the natural history of animal architecture*. Oxford University Press, Oxford.

Hartshorne, C (1958) The relation of bird song to music. *Ibis*, 100, 421–445.

Hendrik, LK et al. (2022) A review of false heads in Lycaenid butterflies. *Journal of the Lepidopterists' Society*, 76, 140–148.

Hinde, RA (ed., 1969) *Bird Vocalizations*. Cambridge University Press, Cambridge.

Hinde, RA (1982) *Ethology*. Fontana, London.

Hinde, RA & Steel, E (1976) The effect of male song on an estrogen-dependent behavior pattern in the female canary (*Serinus canarius*). *Hormones and Behavior*, 7, 293–304.

Hofstadter, DR & Dennett, DC (1981) *The Mind's I*. Harvester Press, Brighton.

Horka, I et al. (2018) Multiple origins and strong phenotypic convergence in fish-cleaning palaemonid shrimp lineages. *Molecular Phylogenetics and Evolution*, 124, 71–81.

Hughes, DP et al. (2012) *Host Manipulation by Parasites*. Oxford University Press, Oxford.

Hull, DL (1981) The units of evolution: a metaphysical essay. In UJ Jensen & R Harré (eds), *The Philosophy of Evolution*. Harvester, London.

Huelsmann, M et al. (2019) Genes lost during the transition from land to water in cetaceans highlight genomic changes associated with aquatic adaptations. *Science Advances*, 5.

Huxley, JS (1923) *Essays of a Biologist*. Chatto & Windus, London.

Huxley, JS (1932) *Problems of Relative Growth*. Dial Press, New York.

Illingworth, S (2020) Tidal evolution. *The Poetry of Science*.

Jablonka, E (2004) From replicators to heritably varying phenotypic traits: the extended phenotype revisited. *Biology and Philosophy*, 19, 353–375.

Joyce, WG & Gauthier, JA (2003) Palaeoecology of Triassic stem turtles sheds new light on turtle origins. *Proc. Roy. Soc. Lond.*, B, 271, 1–5.

Joyce, H (2021) *Trans: when ideology meets reality*. Oneworld, London.

Kaji, T et al. (2018) Parallel saltational evolution of ultrafast movements in snapping shrimp claws. *Current Biology*, 28, 106–113.

Kalluraya, CA et al. (2023) Bacterial origin of a key innovation in the evolution of the vertebrate eye. *Proc. Nat. Acad. Sci.*, 120.

Kettlewell, HBD (1973). *The Evolution of Melanism.* Oxford University Press, Oxford.

Kimura, M (1983) *The Neutral Theory of Molecular Evolution.* Cambridge University Press, Cambridge.

Kingdon, J (2003) *Lowly Origin.* Princeton University Press, Princeton, NJ.

King's College Sociobiology Group (1982). *Current Problems in Sociobiology.* Cambridge University Press, Cambridge.

Konishi, M & Nottebohm, F (1969) Experimental studies in the ontogeny of avian vocalizations. In RA Hinde (ed.), *Bird Vocalizations.* Cambridge University Press, Cambridge.

Kowalczyk, A et al. (2022) Complementary evolution of coding and noncoding sequence underlies mammalian hairlessness, *eLife*, 11, 7 Nov.

Krebs, JR (1977) The significance of song repertoires: the Beau Geste hypothesis. *Animal Behaviour*, 25, 475–478.

Krebs, JR & Davies, NB (eds, 1978, 1984, 1991) *Behavioural Ecology: an evolutionary approach.* Blackwell Scientific Publications, Oxford.

Krebs, JR & Davies, NB (1987) *An Introduction to Behavioural Ecology.* Blackwell Scientific Publications, Oxford.

Krebs, JR & Dawkins, R (1984) Animal signals: mindreading and manipulation. In JR Krebs & NB Davies (eds), *Behavioural Ecology* (Second Edition), Blackwell Scientific Publications, Oxford, 380–402.

Kringelbach, ML & Berridge, KC (2010) The functional anatomy of pleasure and happiness. *Discov. Med.*, 9, 579–587.

Kroodsma, DH & Miller, EH (eds, 1982) *Acoustic Communication in Birds.* Volume 2. Academic Press, New York.

Laland, K (2004) Extending the extended phenotype. *Biology and Philosophy*, 19, 313–325.

Land, MF (1980) Optics and vision in invertebrates. In H Autrum (ed.), *Handbook of Sensory Physiology*, 7, 471–592. Springer-Verlag, Berlin.

Le Boeuf, BJ (1974) Male–male competition and reproductive success in elephant seals. *American Zoologist*, 14, 163–176.

Le Boeuf, B & Reiter, J (1988) Lifetime reproductive success in northern elephant seals. In TH Clutton-Brock (ed.), *Reproductive Success.* Chicago University Press, Chicago, 344–362.

Le Duc, D et al. (2022) Genomic basis for skin phenotype and cold adaptation in the extinct Steller's sea cow. *Science Advances*, 8.

Lehrman, DS (1964) The reproductive behavior of ring doves. *Scientific American*, 211, 48–55.

Leigh, EG (1971) *Adaptation and Diversity.* Freeman, Cooper, San Francisco.

Lents, NH (2019) *Human Errors.* Houghton Mifflin, Boston/New York.

LePage, M (2023) Life-extending parasite makes ants live at least three times longer. *New Scientist*, 12 June.

Lettvin, JY et al. (1959) What the frog's eye tells the frog's brain. *Proceedings of the I.R.E.*, 47, 1940–1951.

Lettvin, JY et al. (1961) Two remarks on the visual system of the frog. In WA Rosenblith (ed.), *Sensory Communication*, MIT Press, Cambridge, MA.

Lewis, CS (1939) Bluspels and flalansferes: a semantic nightmare. In *Rehabilitations and Other Essays*. Oxford University Press, Oxford.

Lewontin, RC (1967) Spoken remark in *Mathematical Challenges to the Neo-Darwinian Interpretation of Evolution*. In PS Morgan & M Kaplan (eds), *Wistar Institute Symposium Monograph*, 5, 79.

Lewontin, RC (1979). Sociobiology as an adaptationist program. *Behavioral Science*, 24, 5–14.

Li, Y et al. (2010) The hearing gene *Prestin* unites echolocating bats and whales. *Current Biology*, 20, 55–56.

Lorenz, K (1966a) *On Aggression*. Methuen, London.

Lorenz, K (1966b) *Evolution and Modification of Behavior*. Methuen, London.

Low, B (2000) *Why Sex Matters*. Princeton University Press, Princeton, NJ.

Luo, K et al. (2019) Novel instance of brood parasitic cuckoo nestlings using bright yellow patches to mimic gapes of host nestlings. *Wilson Journal of Ornithology*, 131, 686–693.

McFarland, D (1985) *Animal Behaviour*. Pitman, London.

Manning, A & Stamp Dawkins, M (1998) *An Introduction to Animal Behaviour*. Cambridge University Press, Cambridge.

Margulis, L (1998). *The Symbiotic Planet*. Weidenfeld & Nicolson, London.

Marler, P & Slabbekoorn, H (eds, 2004) *Nature's Music: the science of birdsong*. Elsevier, Amsterdam.

Martin, JW et al. (eds), (2014) *Atlas of Crustacean Larvae*. Johns Hopkins University Press, Baltimore.

Maynard Smith, J (1978) *The Evolution of Sex*. Cambridge University Press.

Mayr, E (1963) *Animal Species and Evolution*. Harvard University Press, Cambridge, MA.

Meaney, G (2022) *Zoology's Greatest Mystery*. ISBN 9798424725319.

Miller, G (2000) *The Mating Mind*. Heinemann, London.

Mills, RE et al. (2007) Which transposable elements are active in the human genome? *Trends in Genetics*, 23, No 4.

Monroe, JG et al. (2022) Mutation bias reflects natural selection in *Arabidopsis thaliana*. *Nature*, 602, 101–105.

Nagel, T (1974) What is it like to be a bat? *Philosophical Review*, 83, 435–450.

Nikaido, M et al. (1999) Phylogenetic relationships among cetartiodactyls based on insertions of short and long interspersed elements: hippopotamuses are the closest extant relatives of whales. *Proceedings of the National Academy of Sciences*, 96, 10261–10266.

Noble, D (2017) *Dance to the Tune of Life: biological relativity*. Cambridge University Press, Cambridge.

O'Brien, HD et al. (2016) Unexpected convergent evolution of nasal domes between Pleistocene bovids and Cretaceous Hadrosaur dinosaurs. *Current Biology*, 26, 503–508.

Orgel, LE & Crick, FHC (1980) Selfish DNA: the ultimate parasite. *Nature*, 284, 604–607.

Østergaard, KH et al. (2013) Left ventricular morphology of the giraffe heart examined by stereological methods. *Anatomical Record*, 296, 611–621.

Owen, D (1980) *Camouflage and Mimicry*. Oxford University Press, Oxford.

Pääbo, S (2014) *Neanderthal Man: in search of lost genomes*. Basic Books, New York.

Packard, A & Wurtz, M (1994) An octopus, *Ocythoe*, with a swimbladder and triple jets. *Philosophical Transactions of the Royal Society B*, 344, 261–275.

Patek, SN (2015) The most powerful movements in biology. *American Scientist*, 103, 330–337.

Payne, RS (1972) The song of the whale. In P. Marler (ed.), *Marvels of Animal Behavior*, 144–167. National Geographic Society, Washington, D.C.

Payne, RS & McVay, S (1971) Songs of humpback whales. *Science*, 173, 585–597.

Perkins, S (2012) Porcupine quills reveal their prickly secrets. Science.org, 10 Dec.

Poulsen, CB et al. (2018) Does mean arterial blood pressure scale with body mass in mammals? Effects of measurement of blood pressure. *Acta Physiologica*, 222, e13010.

Pratt, HD (2005) *The Hawaiian Honeycreepers*. Oxford University Press, Oxford.

Pray, L & Zhaurova, K (2008) Barbara McClintock and the discovery of jumping genes (transposons). *Nature Education*, 1, 169.

Pride, D (2020) Viruses can help us as well as harm us. *Scientific American*, 323, 6, 46–53.,

Pringle, JWS (1951) On the parallel between learning and evolution. *Behaviour*, 3, 174–214.

Quackenbush, EM (1968) From Sonsorol to Truk: a dialect chain. PhD thesis, University of Michigan.

Radford, C et al. (2020) Artificial eyespots on cattle reduce predation by large carnivores. *Communications Biology*, 3, 430.

Reich, D (2018) *Who We Are and How We Got Here*. Oxford University Press, Oxford.

Ridley, Mark (2001) *The Cooperative Gene* (previously published (2000) in Britain as *Mendel's Demon*). Free Press, New York.

Ridley, Matt (2020) *How Innovation Works*. Fourth Estate, London.

Robinson, JM (2023) *Invisible Friends*. Pelagic, London.

Roeder, KD & Treat, A (1961) The detection and evasion of bats by moths. *American Scientist*, 49, 135–148.

Romer, AS (1933) *Man and the Vertebrates*. Volume 1. Reprinted by Penguin, London, 1954.

Rowland, T (2010) Running with the grunion. *Santa Barbara Independent*, 9 April.

Ruse, M (2010) Evolution and the idea of social progress. In DR Alexander & RL Numbers (eds), *Biology and Ideology from Descartes to Dawkins*. Chicago University Press, Chicago.

Sandkam, BA (2021) Extreme Y chromosome polymorphism corresponds to five male reproductive morphs of a freshwater fish. *Nature, Ecology and Evolution*, 5, 939–948.

Scally, A (2012) What have we got in common with a gorilla? *Sanger Institute Press Release*, 7 March.

Schluter, D & McPhail, JD (1992) Ecological character displacement and speciation in sticklebacks. *American Naturalist*, 140, 85–108.

Shaffner, SF (2004) The X chromosome in population genetics. *Nature Reviews (Genetics)*, 5, 43–51.

Sheldrake, M (2020) *Entangled Life*. Penguin, London.

Sheppard, PM (1975) *Natural Selection and Heredity*. Hutchinson, London.

Shubin, N (2020) *Some Assembly Required*. Pantheon, New York.

Simon, M (2014) Absurd creature of the week: the parasitic worm that turns snails into disco zombies. *Wired*, 18 Sept.

Simpson, GG (1953) *The Major Features of Evolution*. Simon & Schuster, New York.

Simpson, GG (1980) *Splendid Isolation*. Yale University Press, New Haven, CT.

Singer, C (1931) *A Short History of Biology*. Oxford University Press, Oxford.

Skelhorn, J. et al. (2010) Masquerade: camouflage without crypsis. *Science*, 327, 51.

Skinner, BF (1984) The phylogeny and ontogeny of behavior. *Behavioral and Brain Sciences*, 7, 669–677.

Smith, JLB (1956) *Old Fourlegs*. Longmans, Green, London.

Sober, E & Wilson, DS (1998) *Unto Others: the evolution and psychology of unselfish behavior*. Harvard University Press, Cambridge, MA.

Spottiswoode, C et al. (2011) Ancient host specificity within a single species of brood parasitic bird. *Proceedings of the National Academy of Sciences*, 108, 17738–17742.

Spottiswoode, C et al. (2022) Genetic architecture facilitates then constrains adaptation in a host–parasite coevolutionary arms race. *Proceedings of the National Academy of Sciences*, 119.

Sterelny, K (2001) *Dawkins vs. Gould: survival of the fittest*. Icon, Cambridge.

Sterelny, K & Kitcher, P (1988) The return of the gene. *The Journal of Philosophy*, 85, 339–361.

Stevenson, J (1969) Song as a reinforcer. In RA Hinde (ed.), *Bird Vocalizations*. Cambridge University Press, Cambridge.

Strömberg, CAE (2006) Evolution of hypsodonty in equids: testing a hypothesis of adaptation. *Paleobiology*, 32, 236–258.

Sykes, B (2001) *The Seven Daughters of Eve*. Bantam Press, London.

Symons, D (1979) *The Evolution of Human Sexuality*. Oxford University Press, New York.

Taborsky, M et al. (2021) *The Evolution of Social Behaviour*. Cambridge University Press, Cambridge.

Tanaka, KD et al. (2005) Yellow wing-patch of a nestling Horsfield's hawk cuckoo *Cuculus fugax* induces miscognition by host: mimicking a gape? *Journal of Avian Biology*, 36, 461–464.

Teeling, EC et al. (2000) Molecular evidence regarding the origin of echolocation and flight in bats. *Nature*, 403, 188–192.

Tenaza, RR (1975) Pangolins rolling away from predation risks. *Journal of Mammalogy*, 56, 257.

Thomas, L (1974) *The Lives of a Cell*. Futura, London.

Thompson, D'Arcy W (1942) *On Growth and Form*. Cambridge University Press, Cambridge.

Thorpe, WH (1961) *Bird Song*. Cambridge University Press, Cambridge.

Tinbergen, N (1951) *The Study of Instinct*. Oxford University Press, Oxford.

Tinbergen, N (1964) On adaptive radiation in gulls. *Zoologische Mededelingen*, 39, 209–223.

Tinbergen, N (1966) *Animal Behavior*. Time, New York.

Tov, E (undated) The Torah Scroll: how the copying process became sacred. TheTorah.com.

Trivers, RL (1972) Parental investment and sexual selection. In B Campbell (ed.), *Sexual Selection and the Descent of Man*. Aldine, Chicago.

Trivers, RL (1985) *Social Evolution*. Benjamin/Cummings, Menlo Park, CA.

Trivers, RL (2000) In memory of Bill Hamilton. *Nature*, 404, 828.

Trivers, RL (2011) *The Folly of Fools*. Basic Books, New York.

Turner, JS (2004) Extended phenotypes and extended organisms. *Biology and Philosophy*, 19, 327–352.

Tyson, N deGrasse (2014) *The Pluto Files*. WW Norton, New York.

Van der Linden, A (2016) No teeth, long tongue, no problem: adaptations for ant-eating. *That's Life*. thatslifesci.com.

Villarreal, LP (2016) Viruses and the placenta: the essential virus first view. *APMIS*, 124, 20–39.

Von Holst, E & von Saint Paul, U (1962) Electrically controlled behavior. *Scientific American*, 236, 50–59.

Waddington, CH (1942) The epigenotype. *Endeavour*, 1, 18–20.

Waddington, CH (1977) *Tools for Thought*. Jonathan Cape, London.

Wedel, MJ (2012) A monument of inefficiency: the presumed course of the recurrent laryngeal nerve in Sauropod dinosaurs. *Acta Palaeontologica Polonica*, 57, 251–256.

Weiner, J (1994) *The Beak of the Finch*. Jonathan Cape, London.

West, G (2017) *Scale*. Penguin, London.

Wickler, W (1968) *Mimicry in Plants and Animals*. Weidenfeld & Nicolson, London.

Williams, GC (1992) *Natural Selection: domains, levels and challenges*. Oxford University Press, New York.

Williams, GC (1966, reprinted 1996a) *Adaptation and Natural Selection*. Princeton University Press, Princeton, NJ.

Williams, GC (1996b) *Plan and Purpose in Nature*. Weidenfeld & Nicolson, London.

Wilson, EO (1975) *Sociobiology: the new synthesis*. Harvard University Press, Cambridge, MA.

Workman, L & Reader, L (2004) *Evolutionary Psychology: an introduction*. Cambridge University Press, Cambridge.

Wrangham, R (2009) *Catching Fire: how cooking made us human*. Profile Books, London.

Wren, PC (1924) *Beau Geste*. Murray, London.

Wynne-Edwards, VC (1962) *Animal Dispersion in Relation to Social Behaviour*. Oliver and Boyd, Edinburgh.

Yanai, I. & Lercher, M (2015) *The Society of Genes*. Harvard University Press, Cambridge, MA.

Yong, E. (2011) Lies, damned lies and honey badgers. *Discover*, 19 Sept.

Zahavi, A & Zahavi, A (1997) *The Handicap Principle: a missing piece of Darwin's puzzle*. Oxford University Press, Oxford.

Zimmer, C (2021) A new company with a wild mission: bring back the woolly mammoth. *New York Times*, 13 Sept.

Image Credits

P 3. Bill Hamilton, reproduced with permission of Dr Mary Bliss
P 4. Minden Pictures / Alamy Stock Photo
P 8. Richard Dawkins
P 9. Richard Dawkins
P 15. Yathin Krishnappa
P 16. Max Allen / Alamy Stock Photo
P 17. Photographer – Michael Carroll / Media Drum World / Alamy Stock Photo
P 18. blickwinkel / Alamy Stock Photo
P 19. (*top*) Bill Coster IN / Alamy Stock Photo
P 19. (*bottom*) Anil Kumar Verma
P 20. Brett Billing and Ryan Hagerty USFWS
P 21. reptiles4all / Shutterstock
P 22. (*top*) yod 67 / Shutterstock
P 22. (*bottom*) André Gilden / Alamy Stock Photo
P 23. (*top*) Jiri Balek / Shutterstock
P 23. (*bottom left*) Professor Michael Sweet
P 23. (*bottom right*) Minden Pictures / Alamy Stock Photo
P 24. HWall / Shutterstock
P 25. (*top right*) Professor Michael Sweet
P 25. (*bottom*) Minden Pictures / Alamy Stock Photo
P 26. (*top*) Super Prin / Shutterstock
P 26. (*bottom*) Azura Ahmad / Alamy Stock Photo
P 27. Cameron Radfords
P 28. (*top*) Alexis Srsa / Shutterstock
P 28. (*bottom left*) Professor Michael Sweet
P 28. (*bottom right*) Husein Latif
P 29. (*top*) 3ffi / Shutterstock
P 29. (*bottom right*) Jamikorn Sooktaramorn / Shutterstock
P 43. Richard Dawkins after Joyce & Gauthier (2003)
P 60. Redrawn from Gregory, RL, 'Mind in Science. A History of Explanations in Psychology and Physics', *Group Analysis: The International Journal of Group-Analytic Psychotherapy* (SAGE Publications, 1983)/© 1983, © SAGE Publications
P 68. Richard Dawkins
P 106. GagliardiPhotography / Shutterstock
P 111. Redrawn from Kowalczyk, A., Chikina, M. and Clark, N. (2022)

'Complementary Evolution of Coding and Noncoding Sequence Underlies Mammalian Hairlessness', eLife 11:e76911

P 114. Cavalli-Sforza, LL and Feldman, MW, *Cultural Transmission and Evolution* (Princeton University Press, 1981)

P 117. Redrawn from Figure 2 in Frolová, P., Horká, I. and Ďuriš, Z. (2022), 'Molecular Phylogeny and Historical Biogeography of Marine Palaemonid Shrimps (Palaemonidae: Palaemonella–Cuapetes group)', Scientific Reports, 12, 15237.

P 129. Redrawn from On Growth and Form by D'Arcy Wentworth Thompson (Cambridge University Press, 1917)

P 131. akg-images / Science Source

P 135. Redrawn from Kalliopi Monoyios in Neil Shubin, *Some Assembly Required: Decoding Four Billion Years of Life, from Ancient Fossils to DNA* (Pantheon, 2020)

P 138. Redrawn from Joel W. Martin, Jørgen Olesen, Jens T. Høeg (eds), *Atlas of Crustacean Larvae* (John Hopkins University, 2014)

P 139. Redrawn from Joel W. Martin, Jørgen Olesen, Jens T. Høeg (eds), *Atlas of Crustacean Larvae* (John Hopkins University, 2014)

P 141. (*top*) Library Book Collection / Alamy Stock Photo

P 141. (*bottom left*) Redrawn from Joel W. Martin, Jørgen Olesen, Jens T. Høeg (eds), *Atlas of Crustacean Larvae* (John Hopkins University, 2014)

P 141. (*bottom right*) Redrawn from Joel W. Martin, Jørgen Olesen, Jens T. Høeg (eds), *Atlas of Crustacean Larvae* (John Hopkins University, 2014), modified after Dahms et al. 2006

P 159. Science History Images / Alamy Stock Photo

P 169. Zoonar GmbH / Alamy Stock Photo

P 170. (*top*) Roger Hanlon

P 170. (*bottom left*) Helmut Corneli / Alamy Stock Photo

P 170. (*bottom right*) FtLaud / Shutterstock

P 173. Bill Waterson / Alamy Stock Photo

P 204. Bentley, D. and Hoy, R. 'The Neurobiology of Cricket Song' (*Scientific American*, 1974)

P 217. Bård G. Stokke, NINA

P 218. Charles Tyler

P 220. Photo by Nick Davies

P 223. (*left*) Photo by W.B. Carr

P 223. (*right*) Rose Thorogood

P 254. Richard Dawkins and Yan Wong, *The Ancestor's Tale: A Pilgrimage to the Dawn of Life* (W&N, 2016)

P 265. Zlir'a / Wikimedia Commons

P 276. Sulston et al. (1983) 'The Embryonic Cell Lineage of the Nematode Caenorhabditis Elegans', Developmental Biology (Elsevier)

P 277. Redrawn from Athena Aktipis, *The Cheating Cell* (Princeton University Press, 2020)

P 282. Wikimedia Commons

P 285. Kalluraya, CA, Weitzel, AJ, Tsu, BV and Daugherty, MD, 'Bacterial Origin of a Key Innovation in the Evolution of the Vertebrate Eye' (PNAS, Vol. 120 | No. 16, Figure 2, A)

P 290. The Reading Room / Alamy Stock Photo

Index

baleen whales 74–5, 83, 107, 120
'barber's pole' badges 115
Barbourofelids 102
Barlow, George: *The Cichlid Fishes* 125
barnacle 140–2, 289–90
Barra, Outer Hebrides 273
bats 7, 73, 94, 98, 103–8, 111, 112, 185
Batty, Wilf 89
bauplan (building plan or body plan) 50–2
bay bugs 133–4, 137
beaks 28, 44, 81–4, 85, 163, 235
'Beau Geste' hypothesis 153–4
beaver 36, 147, 212
bees 95, 96, 142, 150, 287, 296
Beethoven, Ludwig van 175
    Pastoral Symphony 217
Bennet-Clark, Henry 201
Bentley, David 203, 204
Biology and Philosophy 212
birdsong 153, 158–62, 234
bithorax 134
bivalve molluscs 94
black-headed gulls 27, 208
blackbird 234–5
blind watchmaker, natural selection as 280
Bliss, Mary 2
blood pressure 68, 110
blue whale 38, 120
body plan 51, 132–3, 187
body temperature 78, 110
boiled frog, fable of the 233–4
bone 10, 43, 50–1, 65, 78, 109–10, 120, 129, 264, 266
bonobos 255, 256, 262, 263
book about past worlds, reading an animal as a 1–3
bow-coo 160, 208
bower birds 208–11
Brachiosaurus 49
Brain
    eye and 48, 62–3, 94
    learning and 143–4, 147–9, 151, 159–61, 163, 164–5, 167, 169, 171–2, 174, 175

memory and 164–5, 171, 221–2
motor homunculus 158–9
parasite and 287
pelvic 'brain' 67–8
size of 81, 163, 280
virtual reality model of the world 5–9, 48, 103–6
brambling 217–19
Brenner, Sydney 275
brine shrimp 51
broken wing display 113–14
Brontosaurus 66, 68–9
brood parasites 214–15, 238
Brooke, Michael 224, 233
Browne, Sir Thomas 142
Burt, Austin: *Genes in Conflict* 241

C

cactus 278, 279
caddis fly (*Silo pallipes*) 196–9, 203, 205–7
Caenorhabditis elegans 275
Cain, Arthur: 'The Perfection of Animals' 58
California Central Valley 263
calling song 201
camera eyes 17, 66, 94, 103
camouflage 14–31, 62, 63, 88, 113, 121, 169–70, 208
canaries 154, 160, 161, 207, 212
cancer 167, 275–6, 279–82
Canidae 70
carapace 129–30, 133
cardinal 236, 237
Carnegie Mellon University 111
Carnivora 102–3
carnivore 69–70, 72–3, 79, 80, 81, 102–3, 266–7, 274
caterpillar 18, 19, 22, 25–6, 28, 29, 30–1, 84, 137, 198, 219, 223–4, 288
cell division 270, 275–80, 283
centipedes 33, 91, 96, 133, 135
cephalopod 94, 169, 170
cerebral cortex 159–60
chameleon 168–9, 185

Felidae 70
Felids 70
Fisher, RA 245, 272, 274
flight 6, 34, 83, 94, 107, 110
fluke 38–9, 120, 288, 292
flying lemurs 97, 108
flying phalangers 98, 107
flying squirrels 96–7, 98, 99
flatfish 169
Ford, EB 270, 272–4
  Ecological Genetics 270
fossils 10, 13, 32, 39, 42, 44, 66, 69,
  72, 74, 80, 81, 83, 120, 261, 264,
  269
fossoriality, dimension of 110
foster parent 162, 213, 214, 216,
  217, 219–20, 222, 230, 233, 234,
  235, 237, 238
Fourier analysis 7–8
fovea 105
fox 110, 111, 113
frequency-dependent selection 229–30
Friedrich, Prince 251
fruit flies 134, 270

G

Galapagos 35, 40, 46, 47, 84–5, 101,
  124
Galapagos finches 84–5
Galapagos marine iguanas 35, 40, 46,
  101
gametes 123, 262–3, 270, 280–1, 286,
  291, 293
gape 70, 83, 163, 234, 235, 237–8,
  239
gastroliths 81, 83
Gauthier, Jacques 41, 42, 44
gavial 73, 74
genes
  active causes, as 176–7, 183–4,
    191
  backward gene's-eye view of
    evolution 214–39, 240, 250, 252,
    256, 258–9, 260
  balance of tensions 188–90
  bookkeeping metaphor 186
  causal influence of 183–5, 190

companion genes 259, 260–82, 284,
  285, 293, 294, 295, 297
driving genes 281
gene discriminators 110
gene pairs 253–4
gene pool 12–14, 16, 88, 143–5,
  146, 152, 184–5, 190, 195, 258,
  259, 260–1, 263, 264–70, 274,
  281, 296, 297
gene tree 250–4
gene's-eye view of evolution 156–7,
  176, 186, 188, 190–4, 198, 206,
  212, 240, 247, 257, 258, 285, 293
'good genes' 184–5
horizontal gene transfer 286
immortality of 10, 179–83
inactivated genes 108
jumping genes 296
'junk' genes 53
'major genes' 190
phenotype and *see* phenotype
pleiotropic (multiple) effects 59, 190
polygenes 205
pseudogenes 53–4
sex-limited genes 226–7, 241
sex-linked genes 226
supergene 274
genome
  computer disc, as 52–4, 180
  fixed 12
  GWAS (genome-wide association
    study) 108–11
  imprinting 240–1
  sequencing 252–5, 256
  viruses and *see* viruses
geometrid stick caterpillar 19
germline mutations 161, 275, 277
giant anteater (*Myrmecophaga*) 75,
  76, 77
giant Galapagos tortoise 46–7
giant isopod 92
giant moa 83
giant sloth 52
gibbon 212
gills 21, 32, 36–7, 38
giraffe 49, 53, 66, 67–8, 120
gizzard 81, 83
gliding 96–7, 108

About the Author and Illustrator

Richard Dawkins is one of the world's most eminent writers and thinkers. He is the award-winning author of *The Selfish Gene*, *The Blind Watchmaker*, *The God Delusion*, and a string of other bestselling science books. He is also a Fellow of the Royal Society and of the Royal Society of Literature. Dawkins lives in Oxford.

Jana Lenzová, born and raised in Bratislava, Slovakia, is an illustrator, translator, and interpreter. After Jana had been commissioned to translate *The God Delusion* by Richard Dawkins into Slovak, she began contributing to his books as an illustrator.